The Gift of Kings

Loro Piana

The Gift of Kings

THE NOBLEST OF WOOLS

PHOTOGRAPHS BY BRUNA ROTUNNO

SKIRA

The story that this book tells starts with the desire to preserve a gift of nature: the world's finest merino wool. In 1975 Franco Loro Piana advised his sons to keep hold of a few lengths of Tasmanian®, the fabric the company had created a few years previously using 17 micron Australian wool. It was an incredible standard of fineness for that period, achieved by local farmers committed to the pursuit of excellence, passionate and proud of their magnificent wool.

Yet the work involved was challenging and costly, and many farmers were already thinking of abandoning it in favour of more intensive production of decidedly lower quality wool. This is why Franco Loro Piana urged his sons to keep hold of the fabric: he feared that such beautiful wool would soon no longer exist.

That concern turned into a new challenge. Franco's son Pier Luigi, passionate about the same exceptional fibre, began to travel around Australia and New Zealand, meeting the woolgrowers who, like him, were devoted to quality and were ready to continue their outstanding work, keen to see it appreciated and supported. Loro Piana took up the challenge with enthusiasm, promoting high quality wool and endorsing the farmers producing it with an annual award. The Record Bale prize was established in 1997 and goes to the two finest bales of wool produced during the year in Australia and New Zealand, which we have purchased ever since.

The prize is not an end in itself, but a way to encourage woolgrowers to keep pursuing the very highest standards. Creating those record bales entails a great deal of hard work: it means continuously producing ultrafine wool of the utmost quality. An exceptional raw material, unthinkable before those trips to the Southern Hemisphere in the late 1970s, is now a reality and has a name: The Gift of Kings®. In fact, in eighteenth-century Spain merino sheep were used as a royal token to seal alliances and honour pacts, like the pact we have made with the woolgrowers of Australia and New Zealand in the name of a shared quest for excellence, and with our demanding customers, always sensitive to quality.

On the backdrop of wild, magnificent landscapes Bruna Rotunno's images capture the sense of this undertaking and this commitment in the eyes of men and women who are passionate about their work, and in the exquisite fleece of the merino sheep.

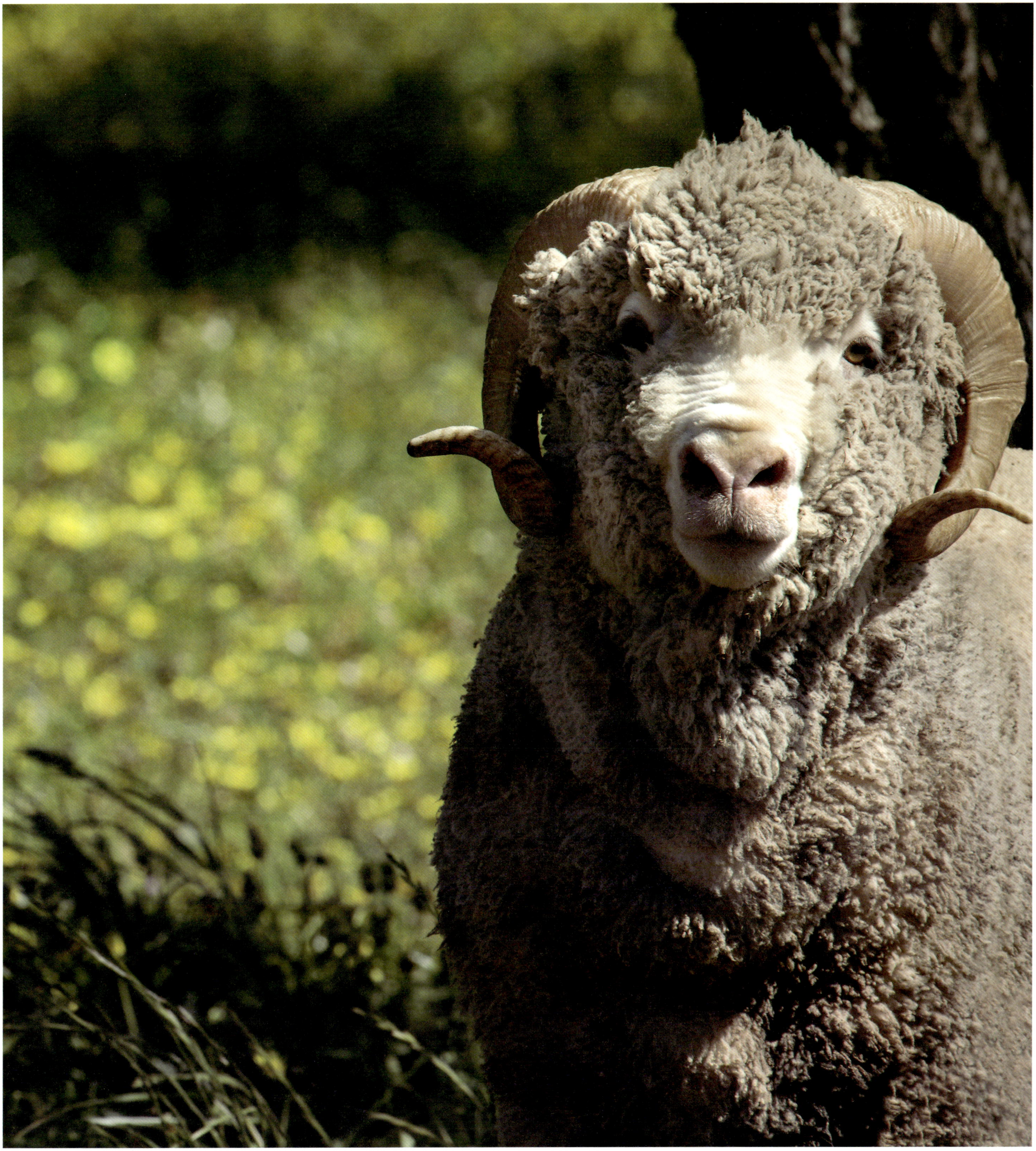

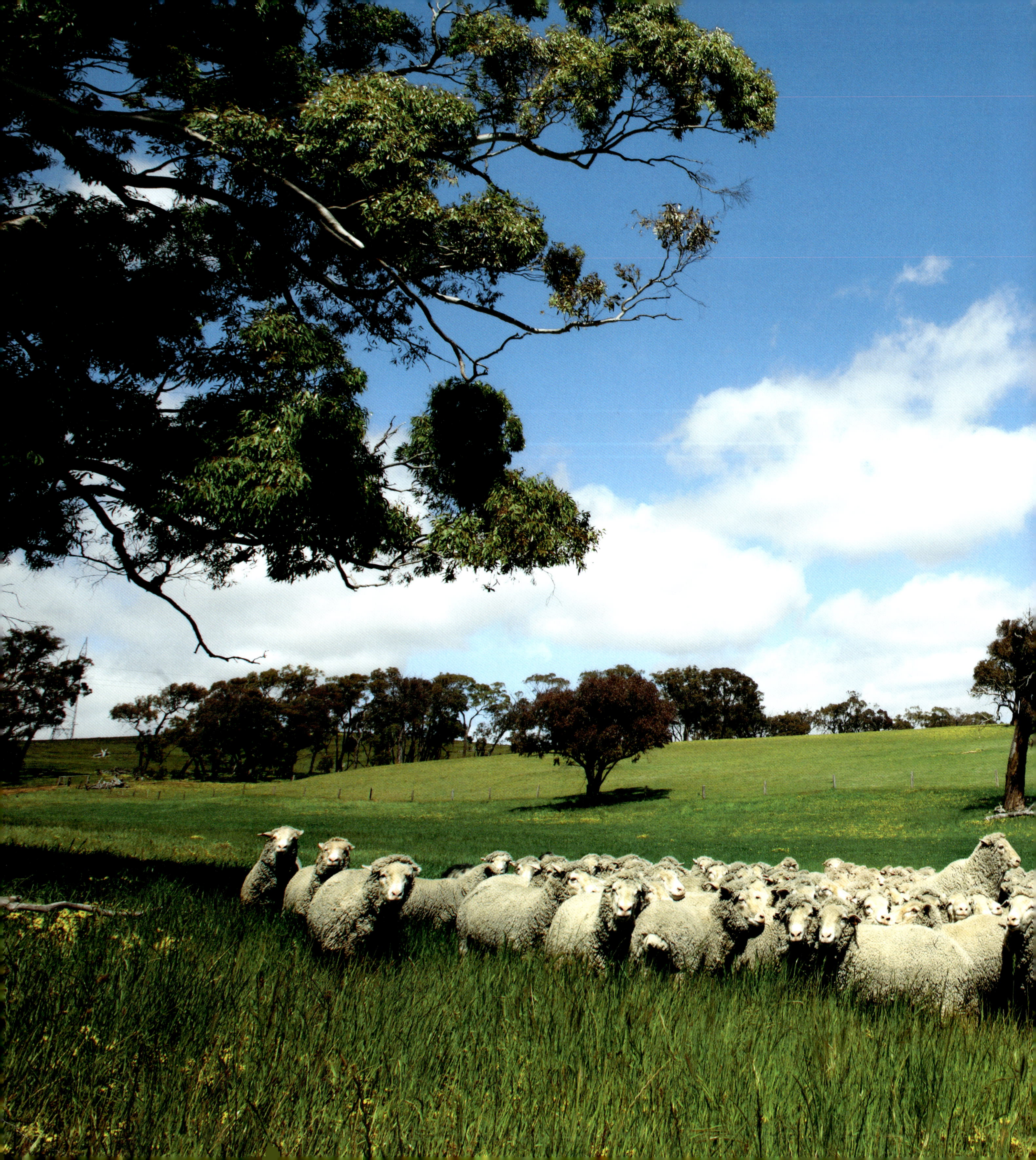

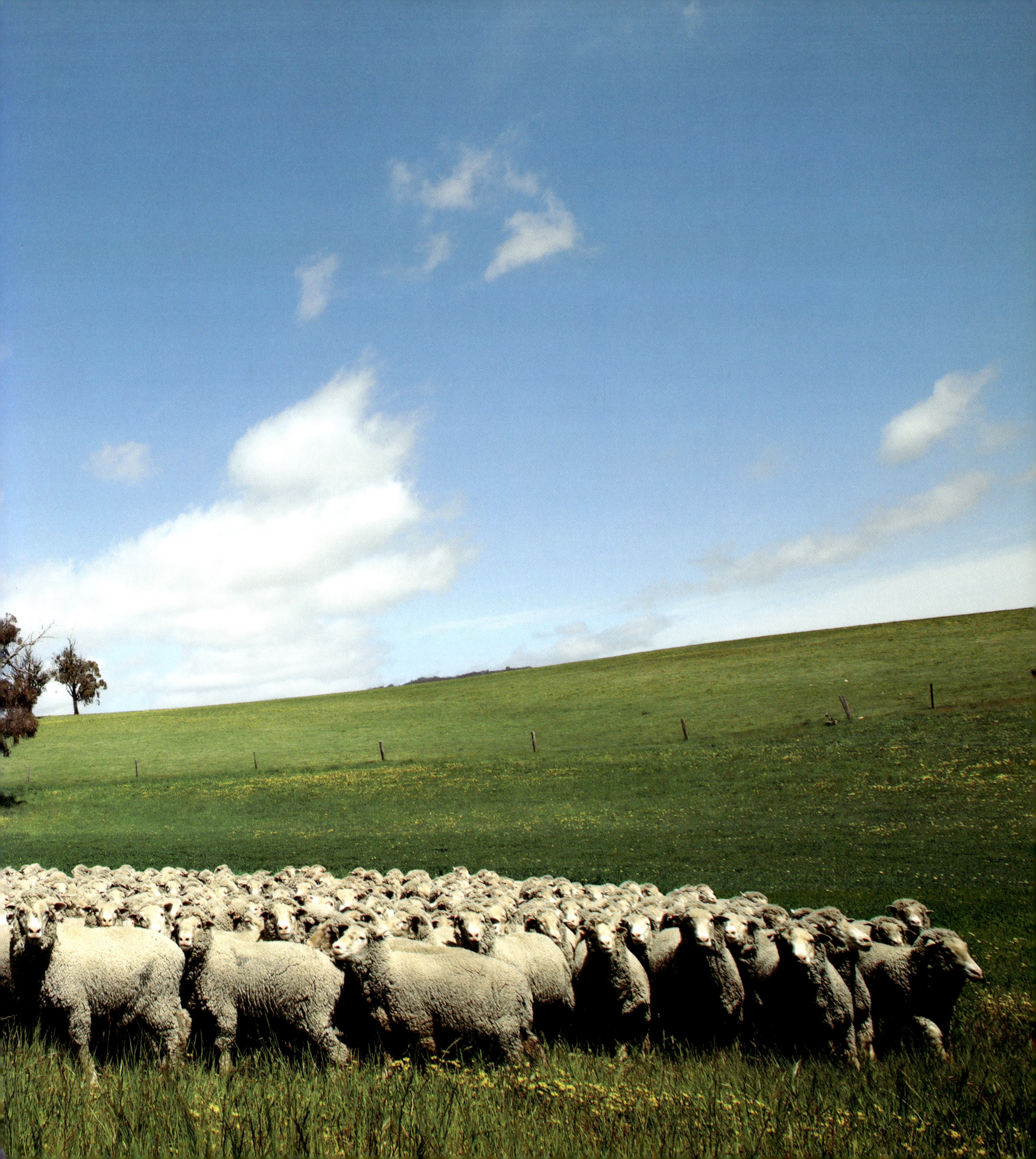

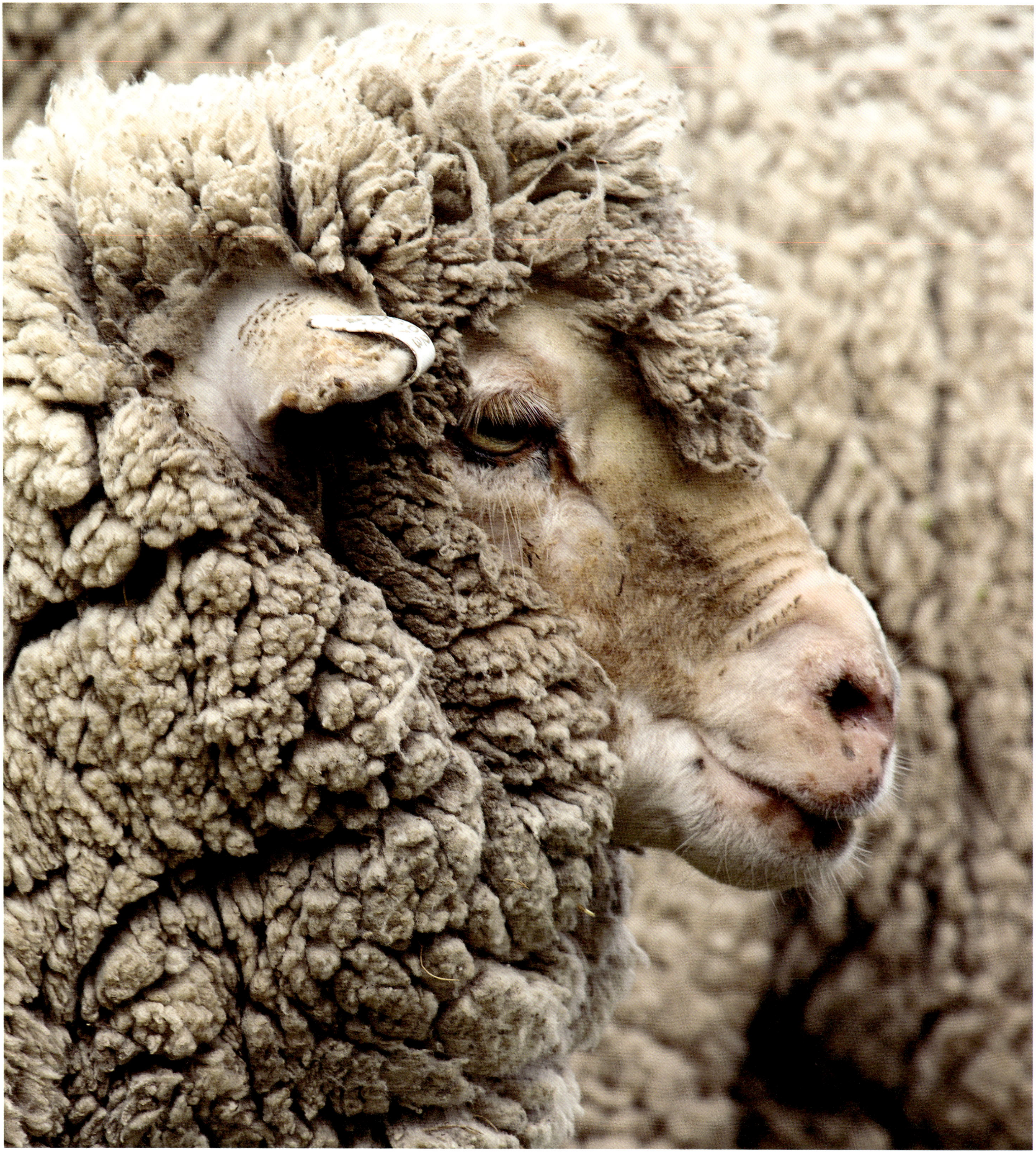

THE FLEECE OF THE MERINO SHEEP TELLS THE STORY OF ITS LIFE, ITS WELFARE. COMPACT, UNIFORM, SOFT, LUSTROUS FIBRE IS THE RESULT OF A HEALTHY EXISTENCE, AT ONE WITH NATURE.

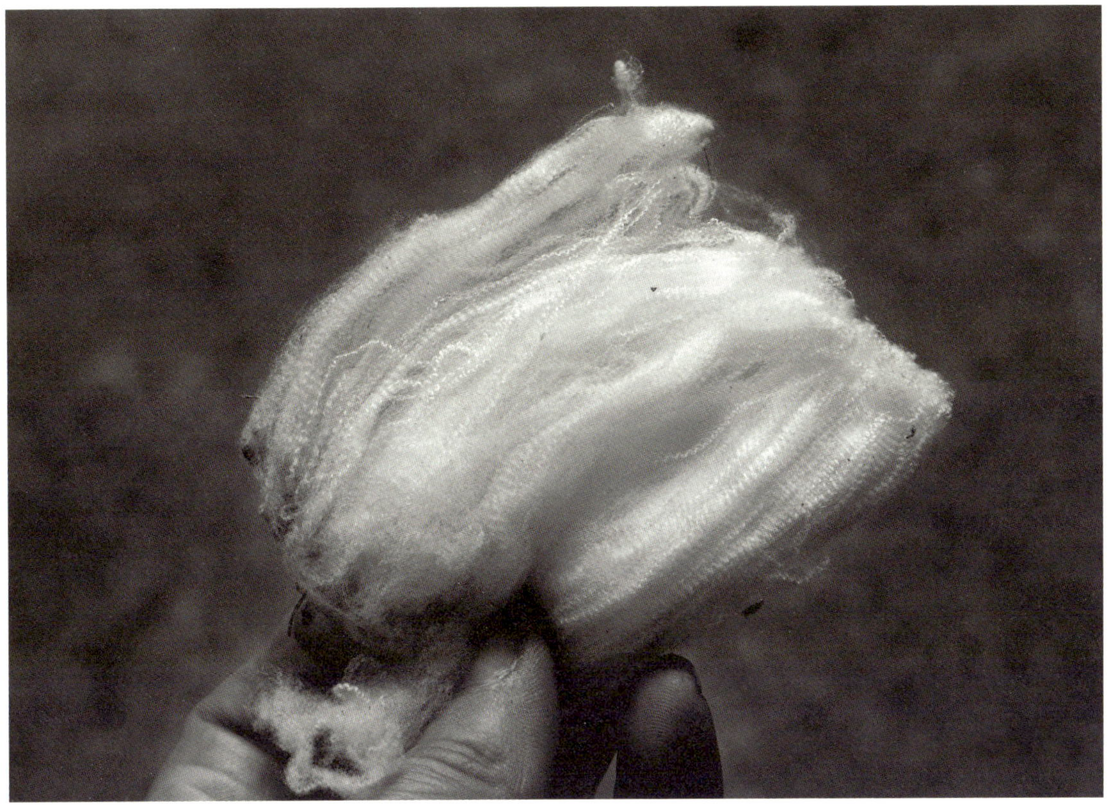

In 1975 Franco Loro Piana took some time out to think about the future of his company. The period had seen some extraordinary achievements, thanks to his tireless quest for the very best. Wool was a genuine passion for him: such a simple, common material, with such a long history, yet capable of radically changing people's lives. The thought that wool used to be considered a basic commodity raises a smile when handling an extrafine merino fabric, like Tasmanian® for example.

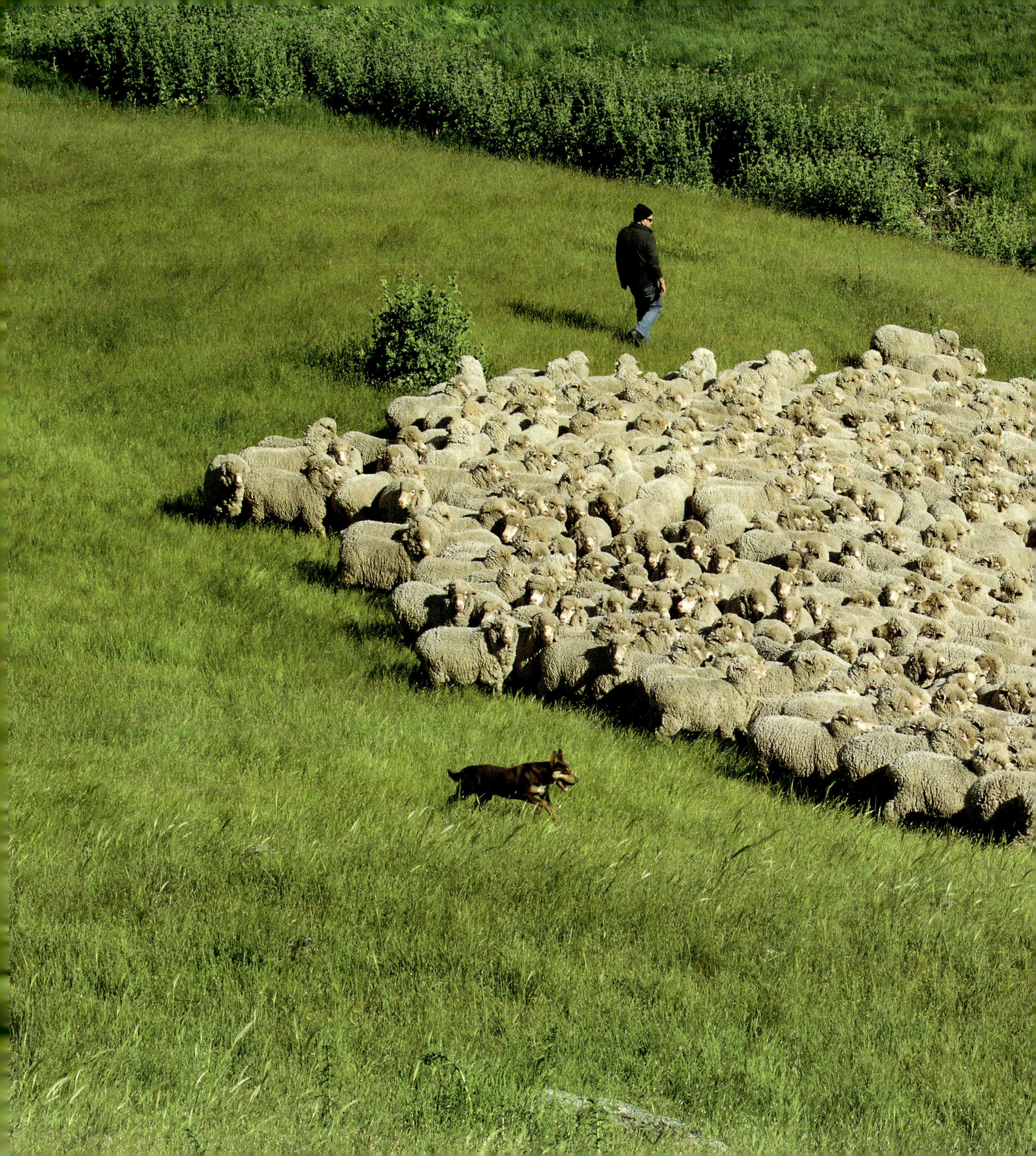

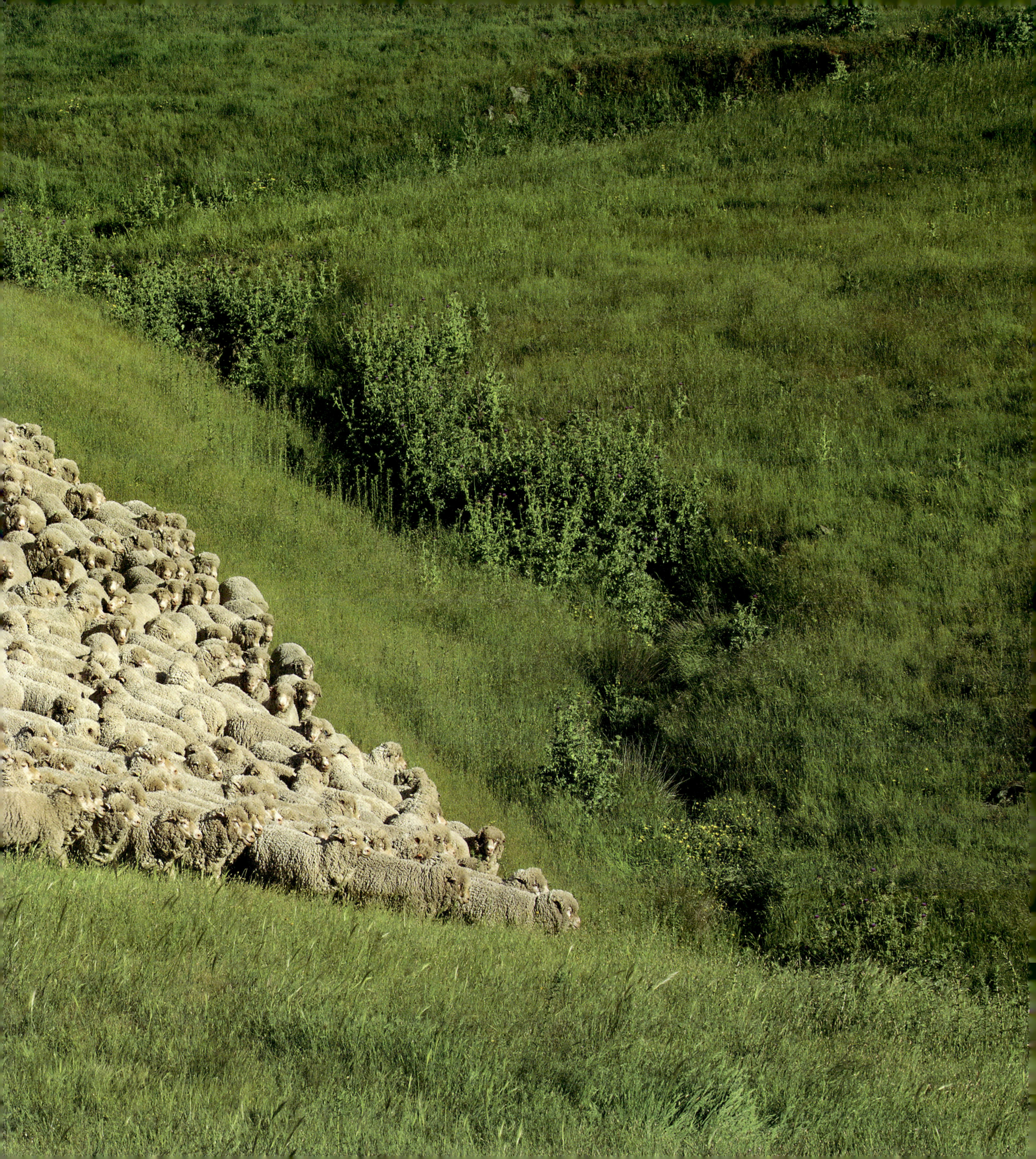

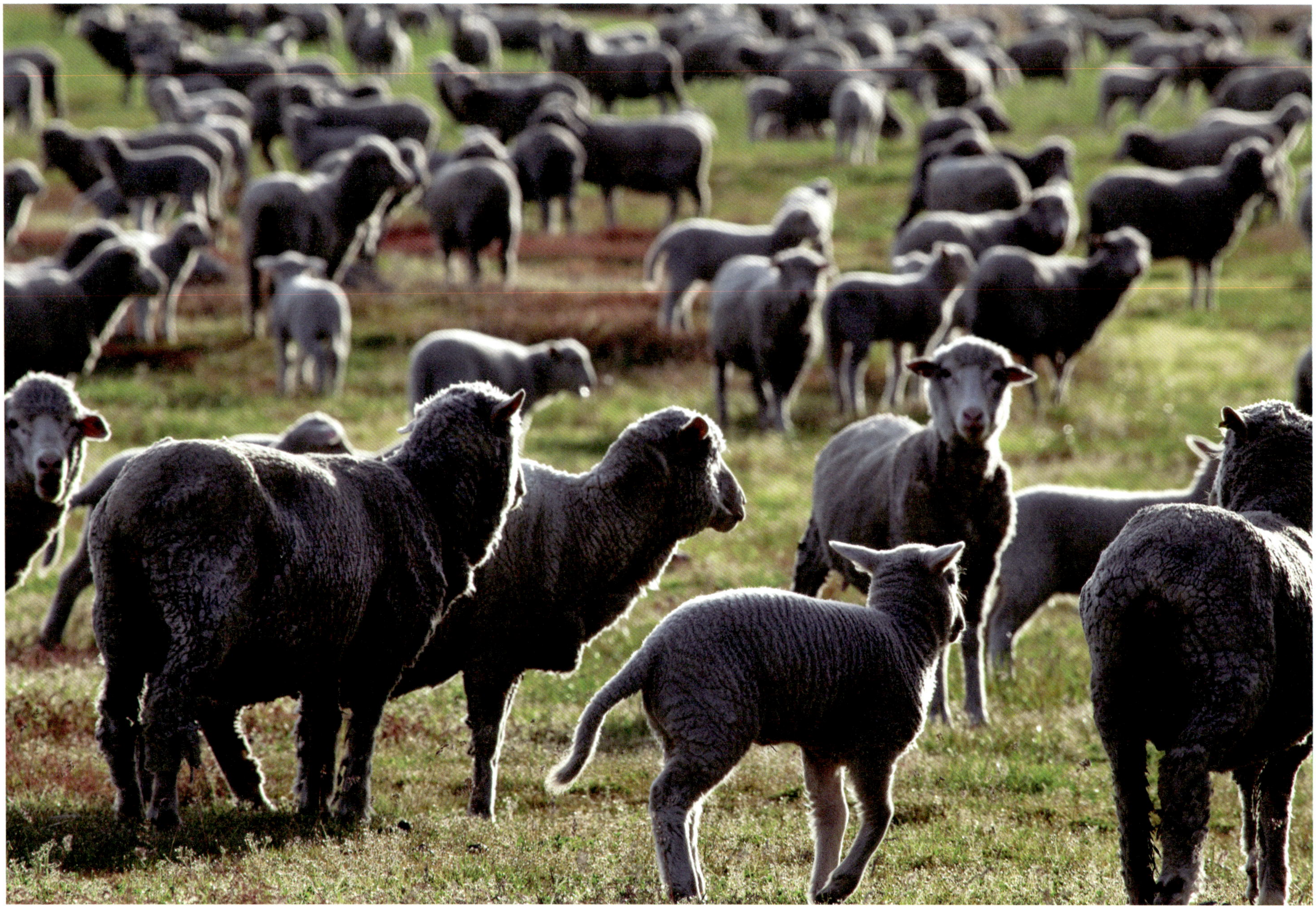

previous pages
At Pyrenees Park, Central Victoria, Australia
the sheep are free to roam, yet always monitored
by the watchful eyes of the growers.
This natural lifestyle guarantees
the welfare of the animals, which produce
exceptional quality wool: light, fine and supple.

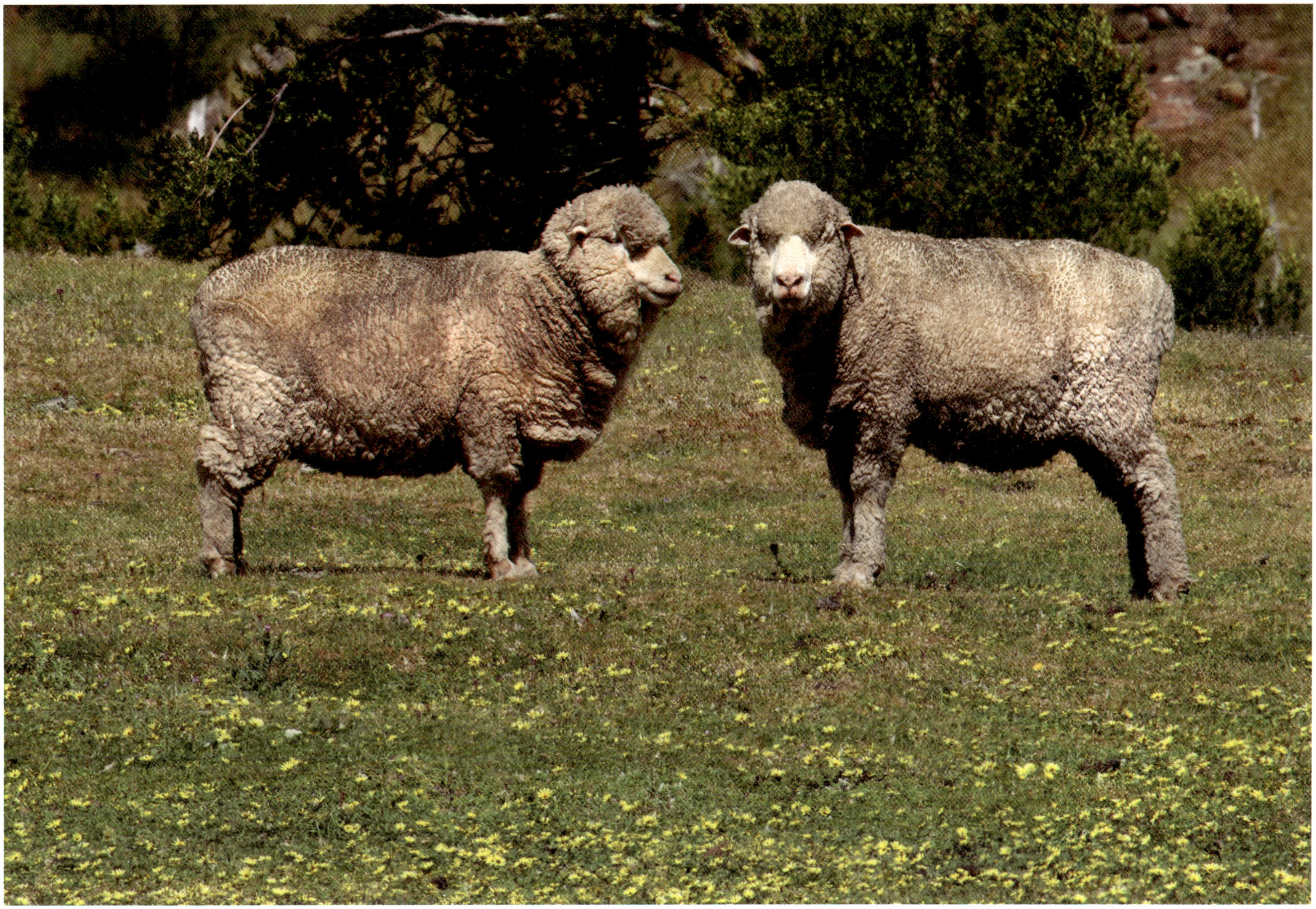

That particular fabric was a genuine milestone for the Italian company. The idea had come to Franco Loro Piana a few years earlier, the starting point being Priest Cloth®, a light-weight fabric produced specifically for clerical clothing. Loro Piana realized that the characteristics of this cloth – fluid, versatile and seasonless – would be ideal for contemporary suiting, crafted from extrafine merino wool sourced from the island of Tasmania, south of Australia. Tasmanian®, a feather-weight worsted fabric that is fine and crease-resistant, perfect for long days in the office, soon became an icon: a symbol of sartorial flair, modernity, comfort and supreme quality.

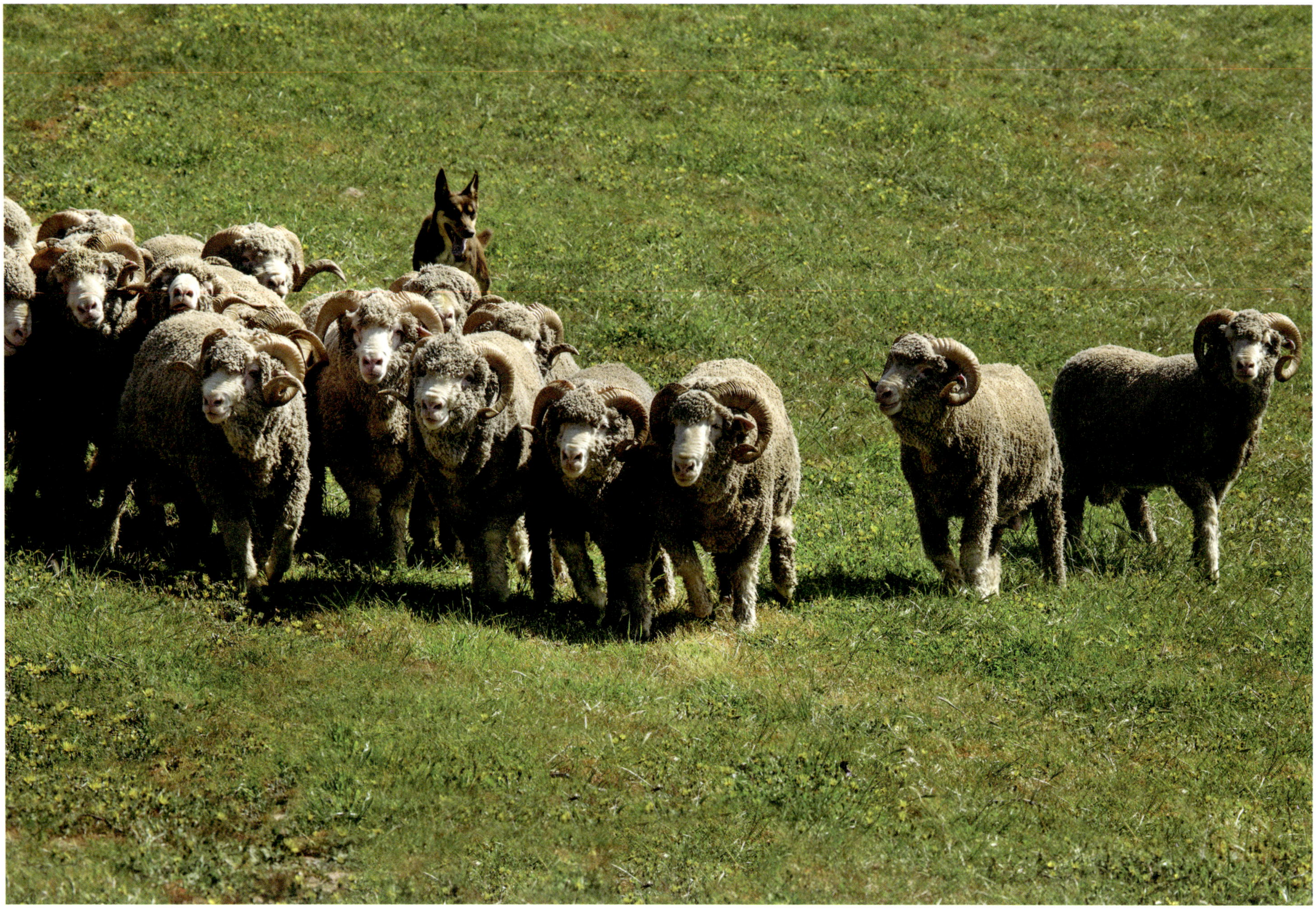

Grazing land in Dunkeld,
Australia.

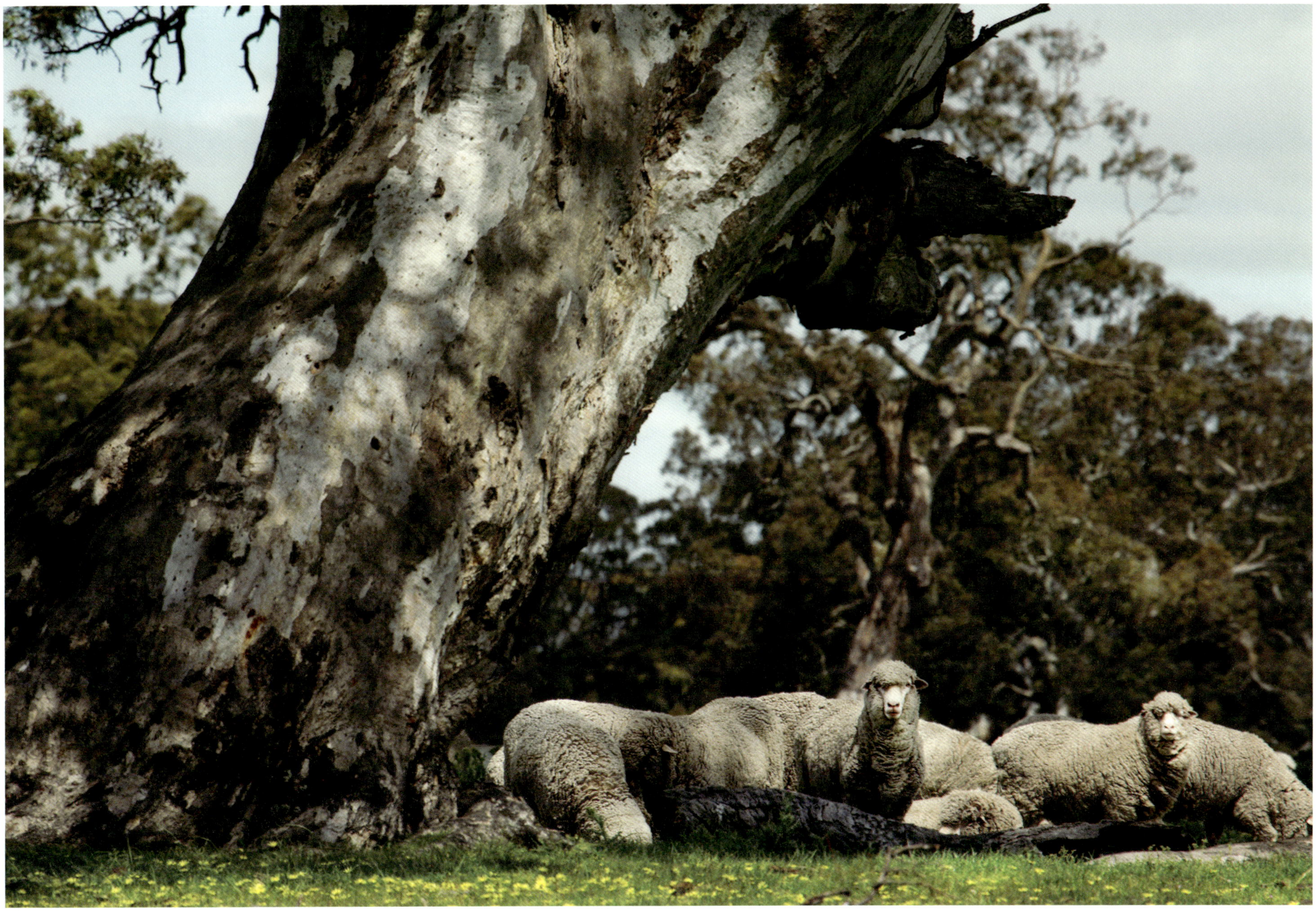

following pages
A strip of land separates the seas
on the coast of Tasmania.

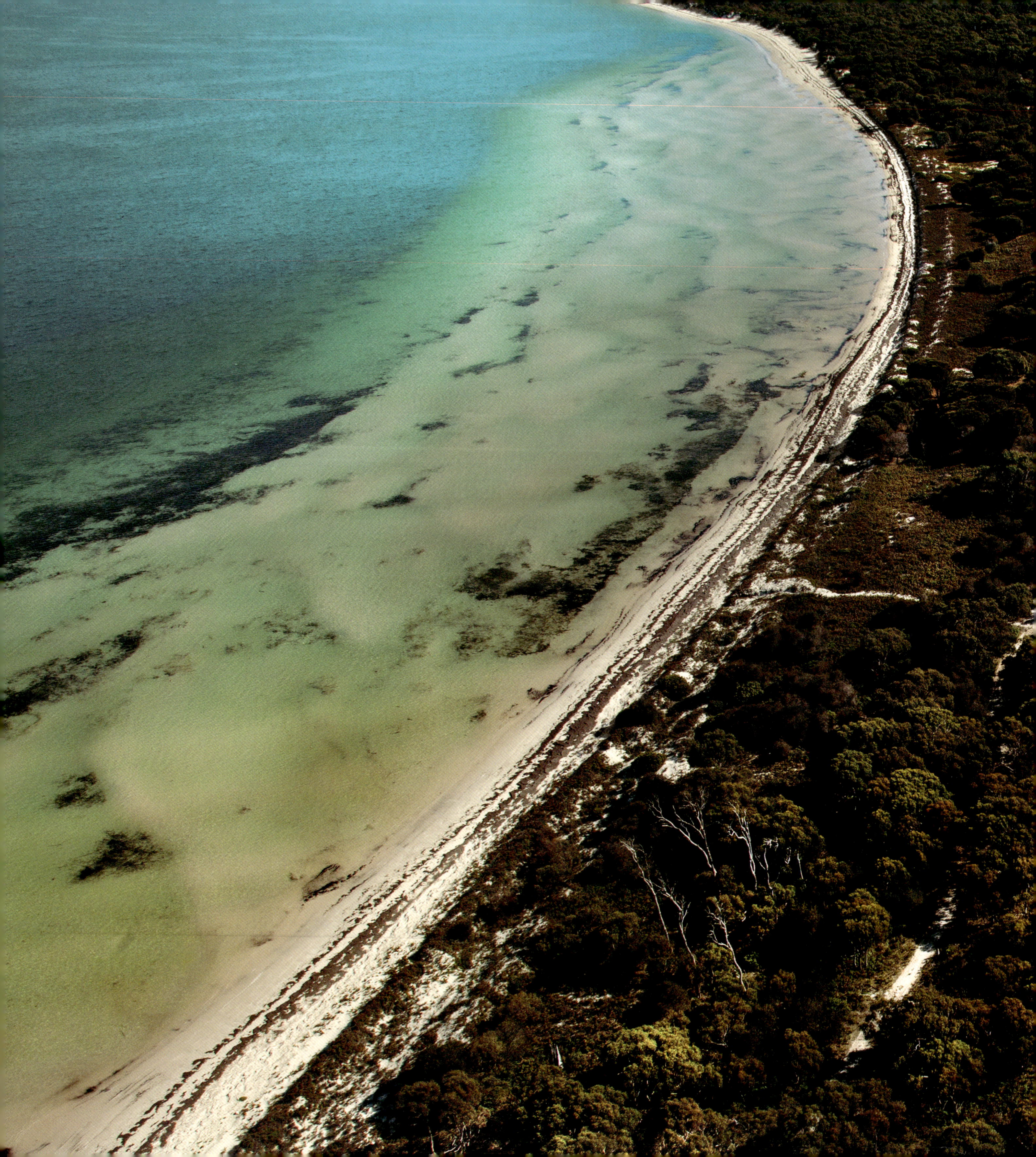

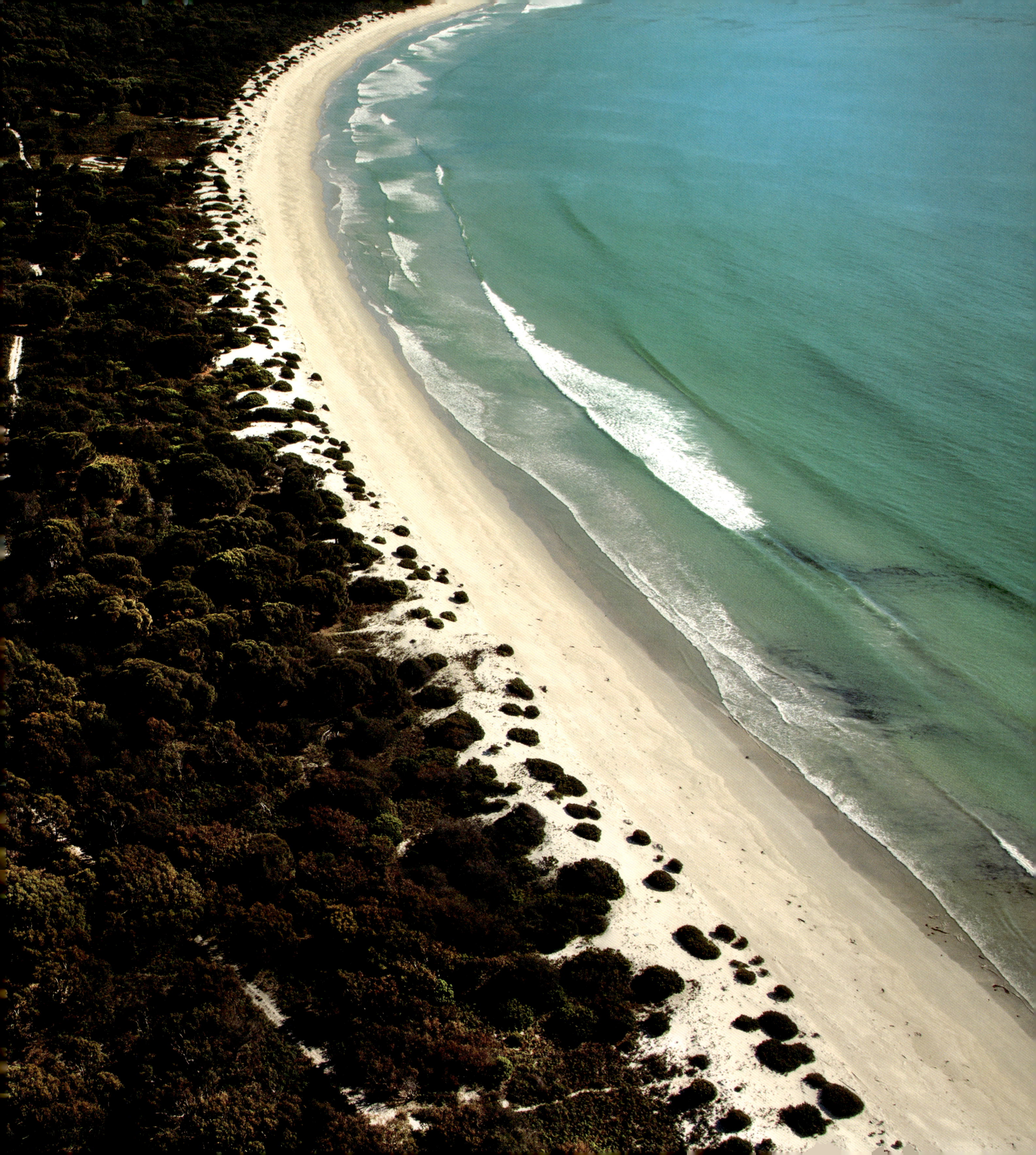

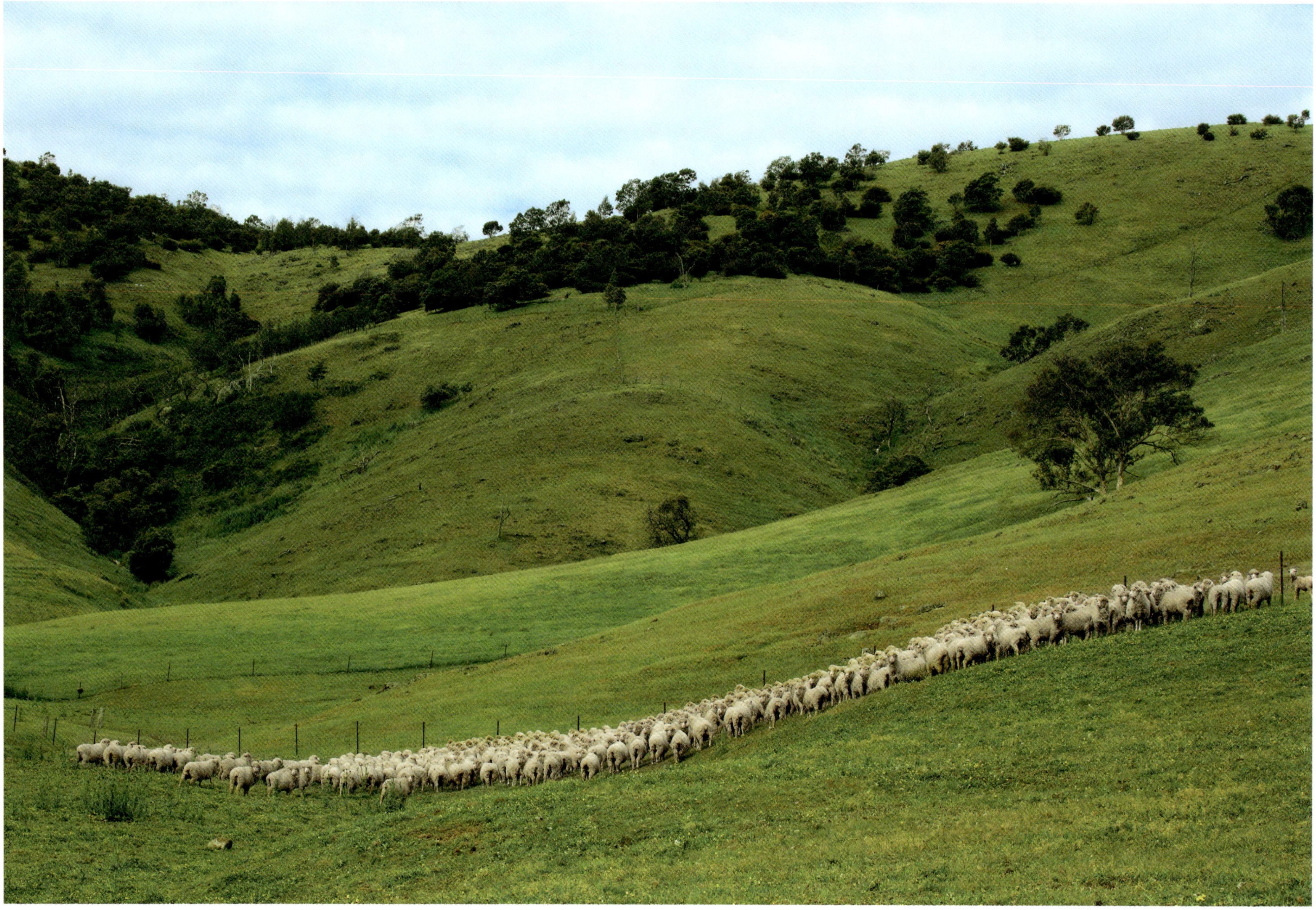

The flocks form white smiles
on the hills in Pyrenees Park farm.

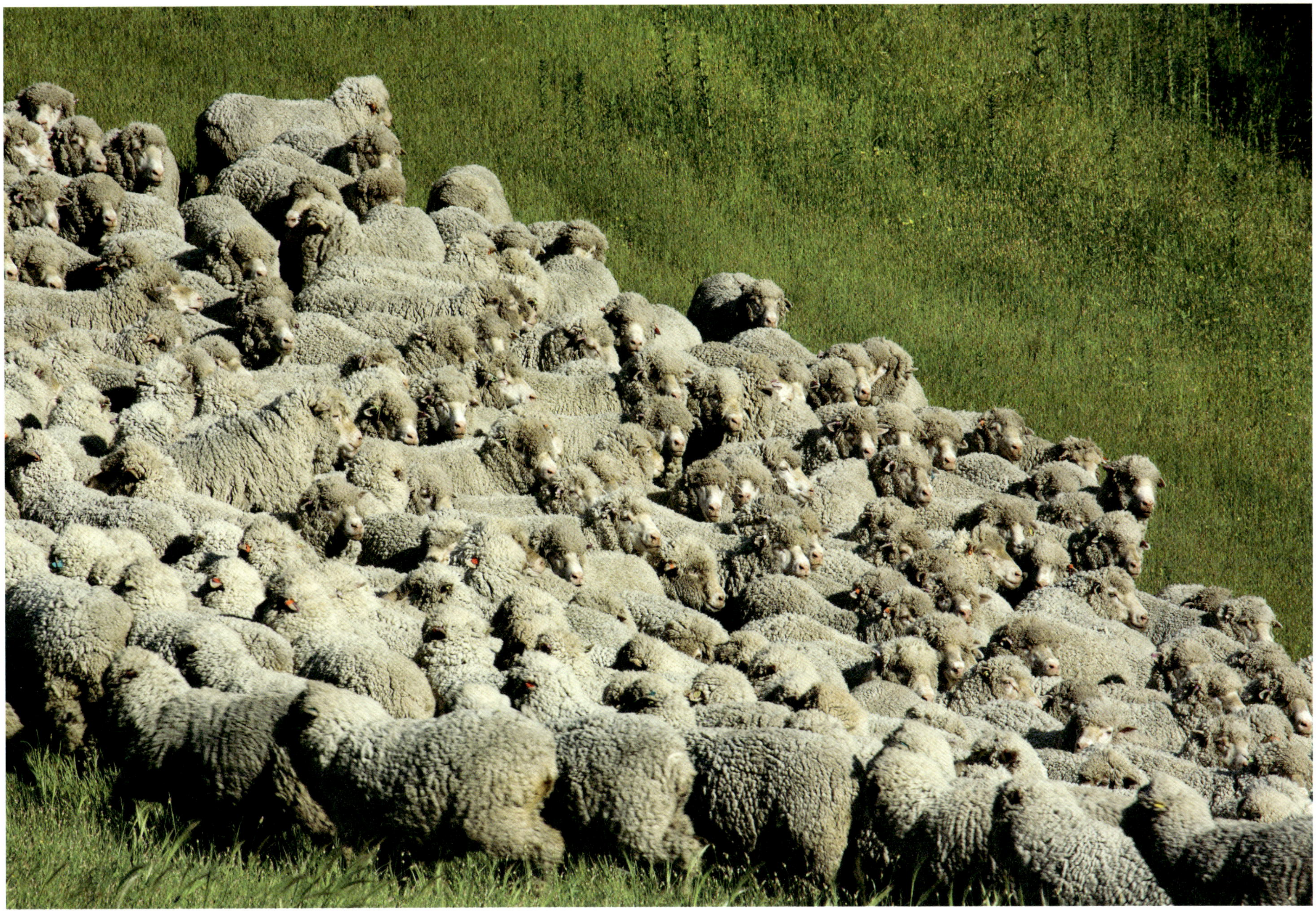

Merino is a special wool, and special care has to be taken of the animals that produce it: they require a suitable environment and experienced growers completely dedicated to this demanding work.

Indeed too demanding, in a market looking for certainties: there were just too many risks involved in such a limited, elite product. Without adequate support, it was clear that the growers would soon abandon these specialized practices for more reliable, profitable wool, destined for a broader market.

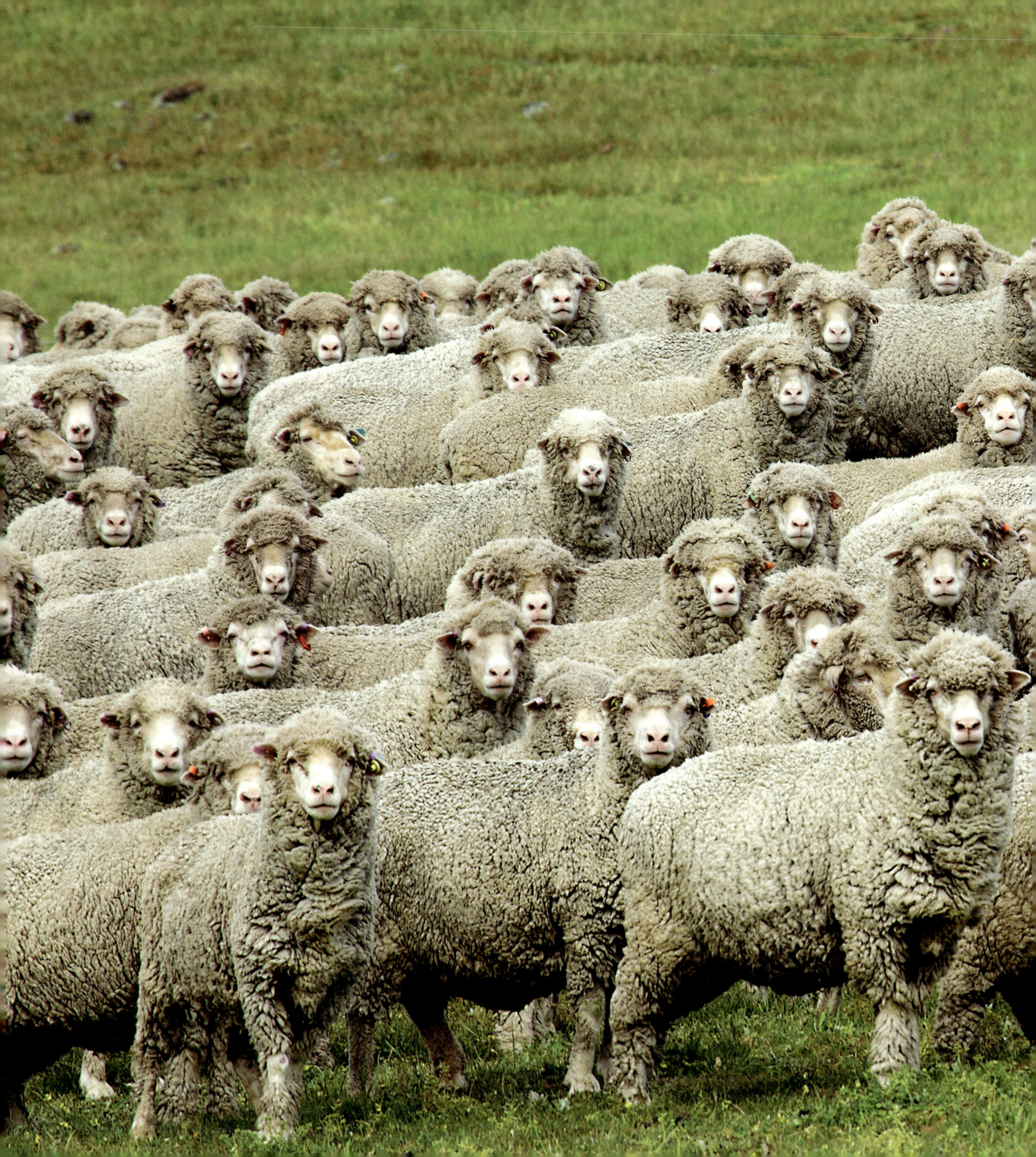

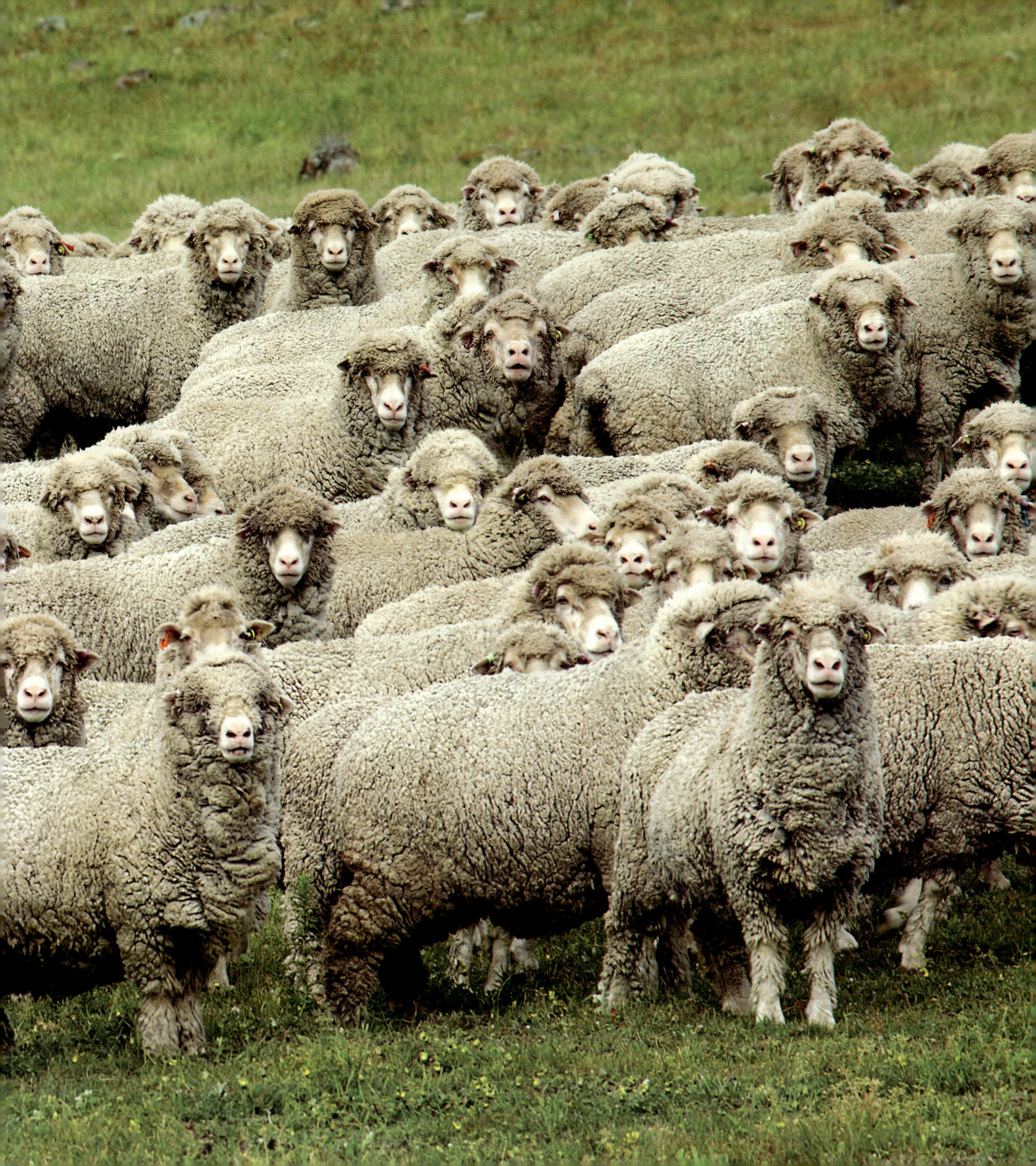

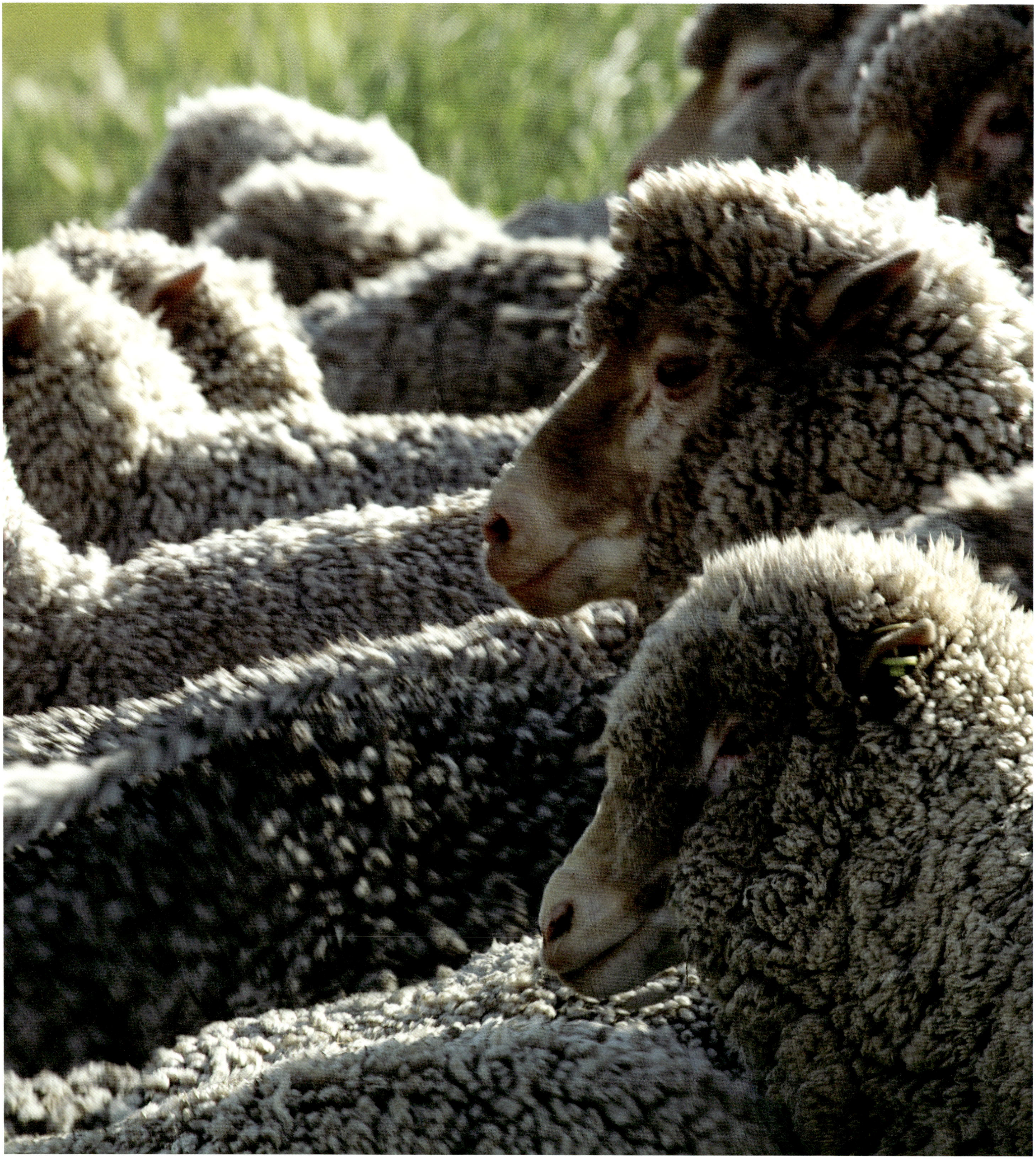

THE ARMIDALE AREA IN AUSTRALIA IS KNOWN FOR ITS NATURAL PARKS, WATERFALLS AND MEADOWS. MERINO WOOL IS A TRULY PRECIOUS COMMODITY, HANDLED WITH CARE AND ATTENTION BY EXPERIENCED GROWERS FOR GENERATIONS.

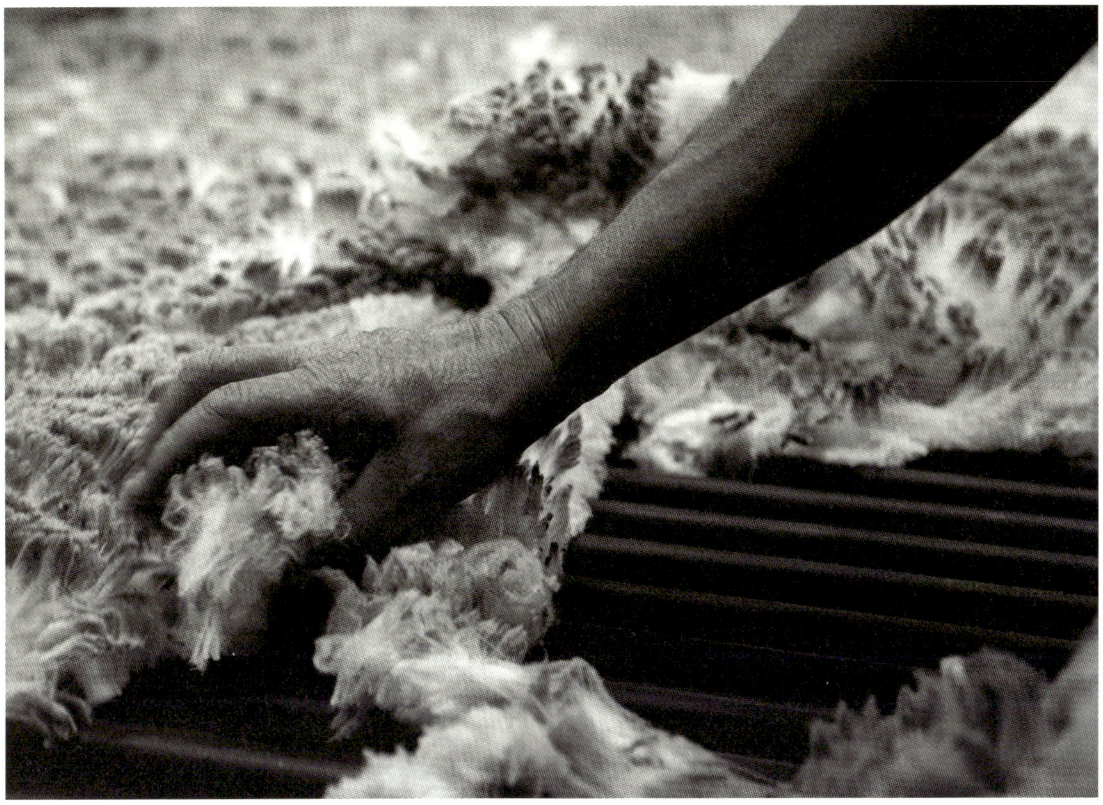

Franco Loro Piana gathered his sons Sergio and Pier Luigi and gave them each a length of that fabric, of that Tasmanian®, synonymous with style and quality, urging them to keep it as a memento of a historical product that would soon no longer exist. That wool was classified as extrafine: it measured 17 microns in diameter.

It is hard to say whether Franco Loro Piana intended his warning as a bleak prediction or a challenge, a stimulus, almost an objective for the future of the company, the passing on of a moral legacy. For Loro Piana, sourcing superlative raw materials is a full-time job, a calling. The thought of losing an excellence was unacceptable.

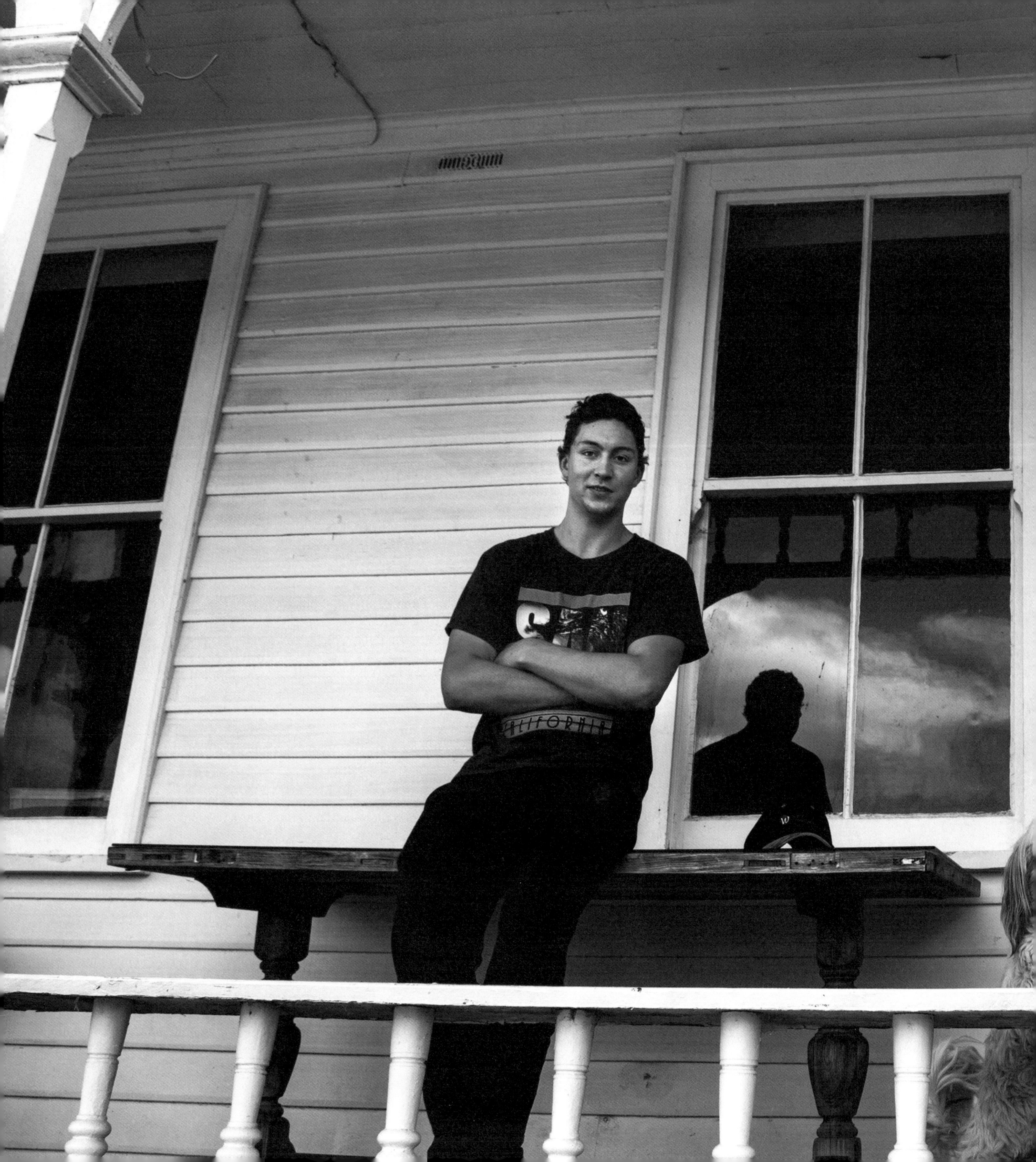

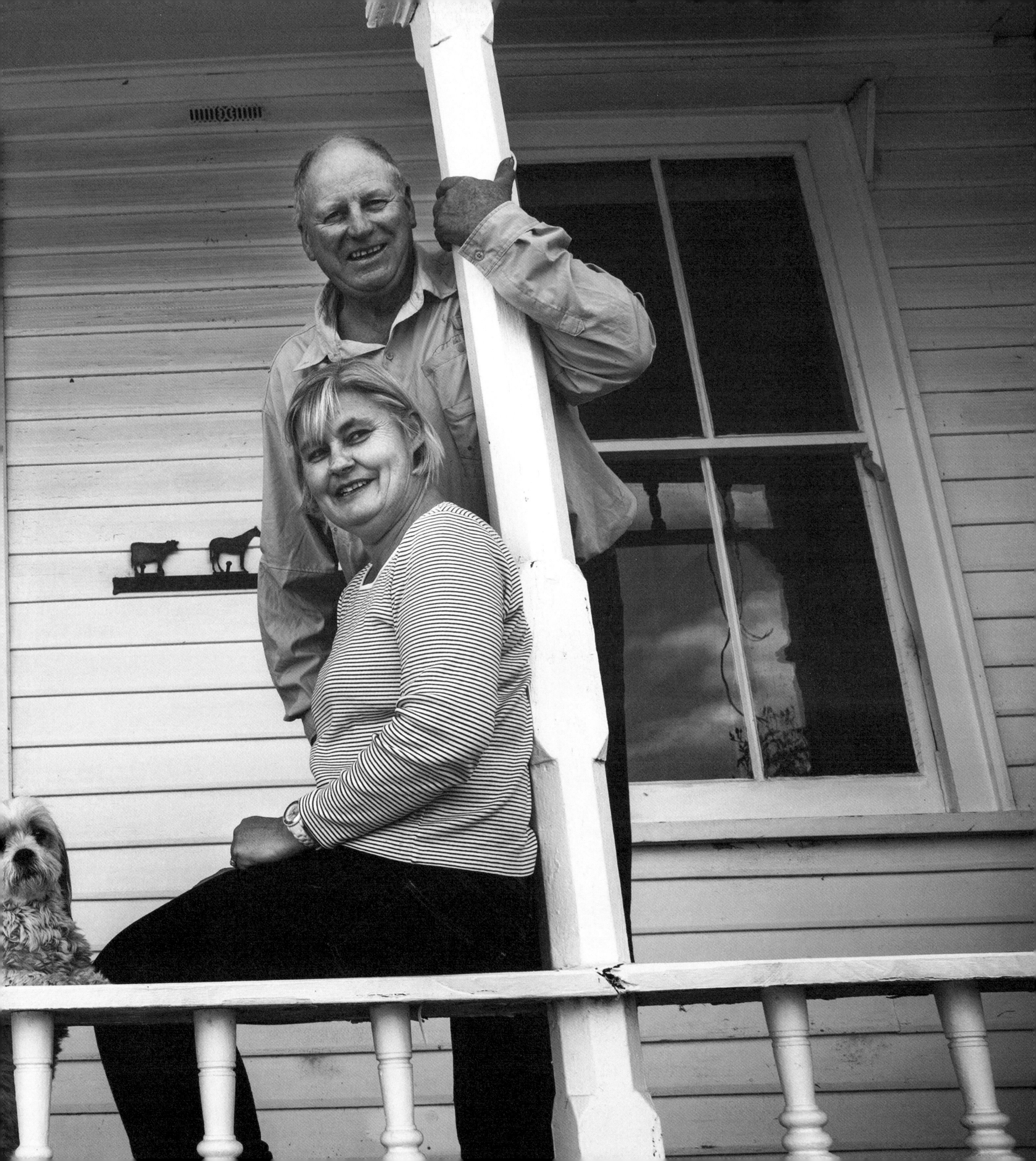

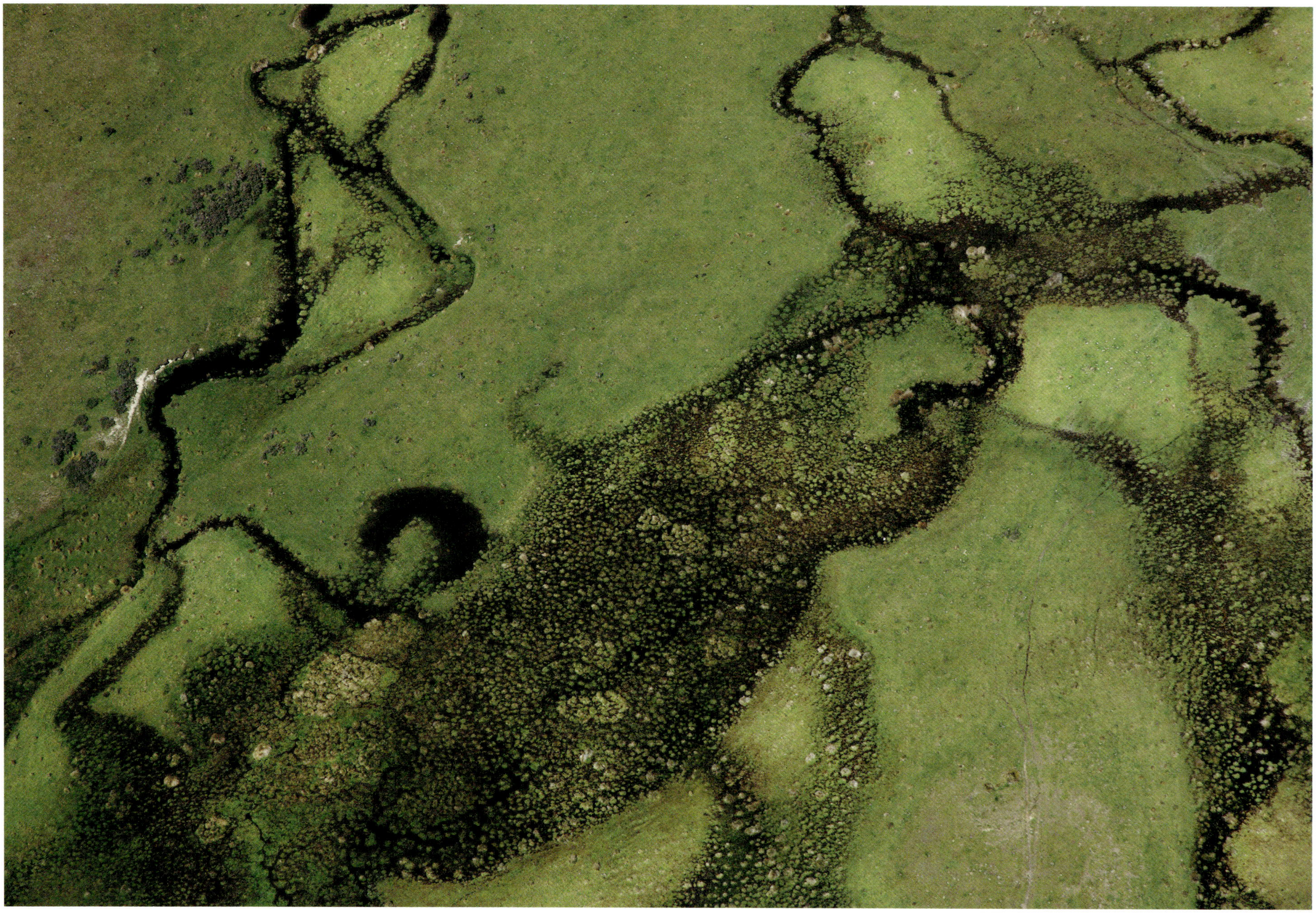

Views of the area around Oatlands,
Tasmania.

previous pages
Mike, Lisa and Hugh Johnston
on their farm in Glenora, Tasmania.

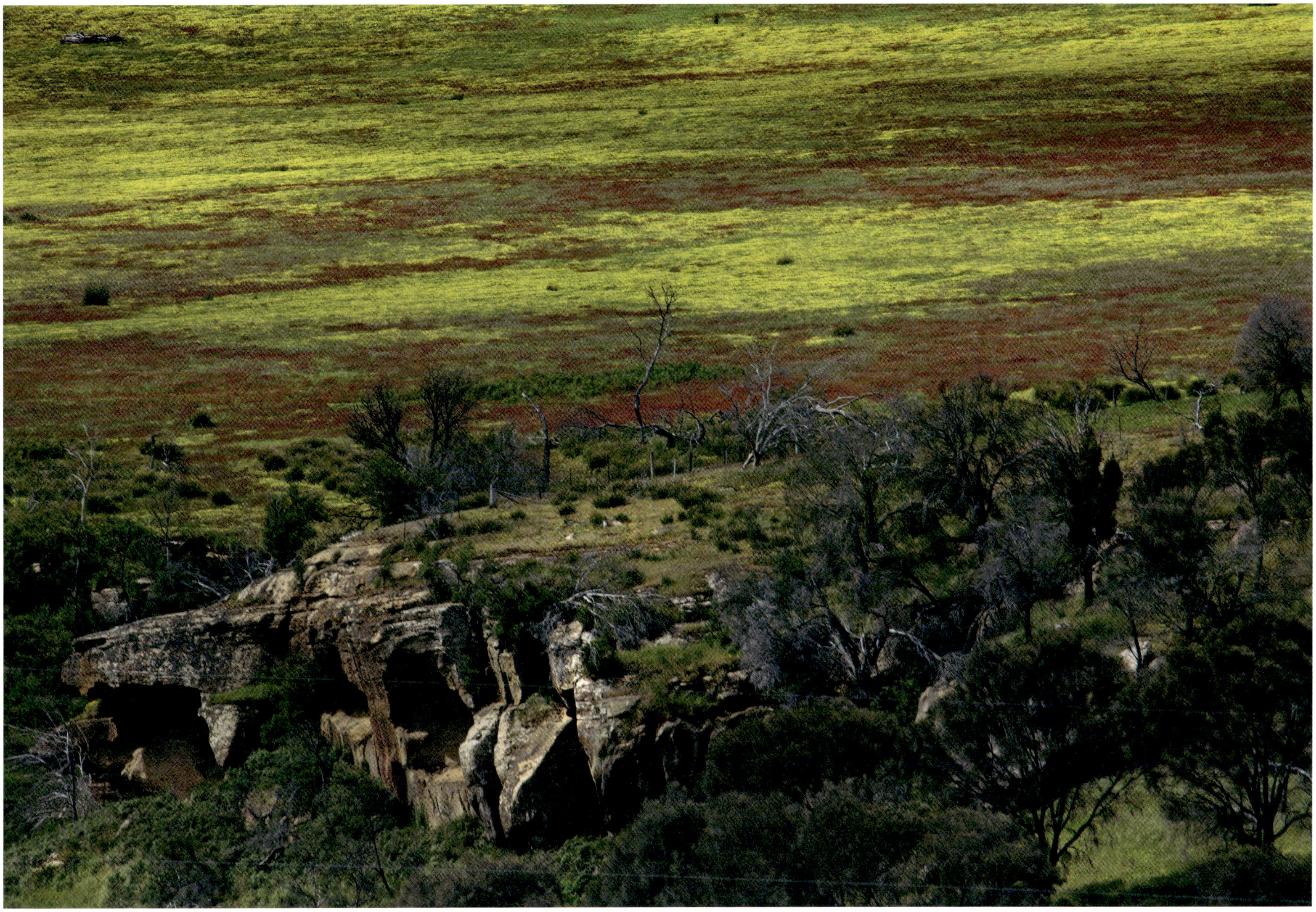

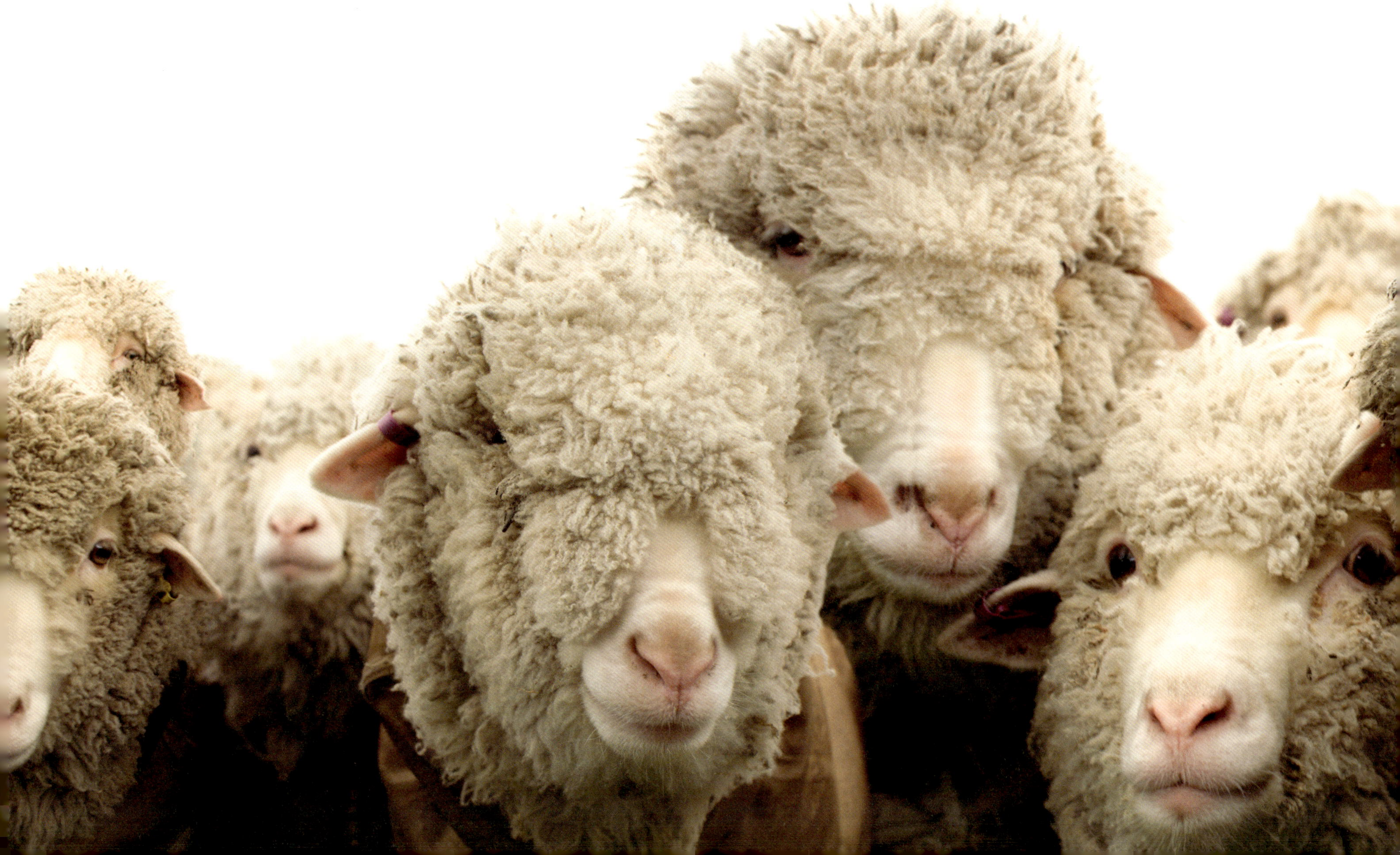

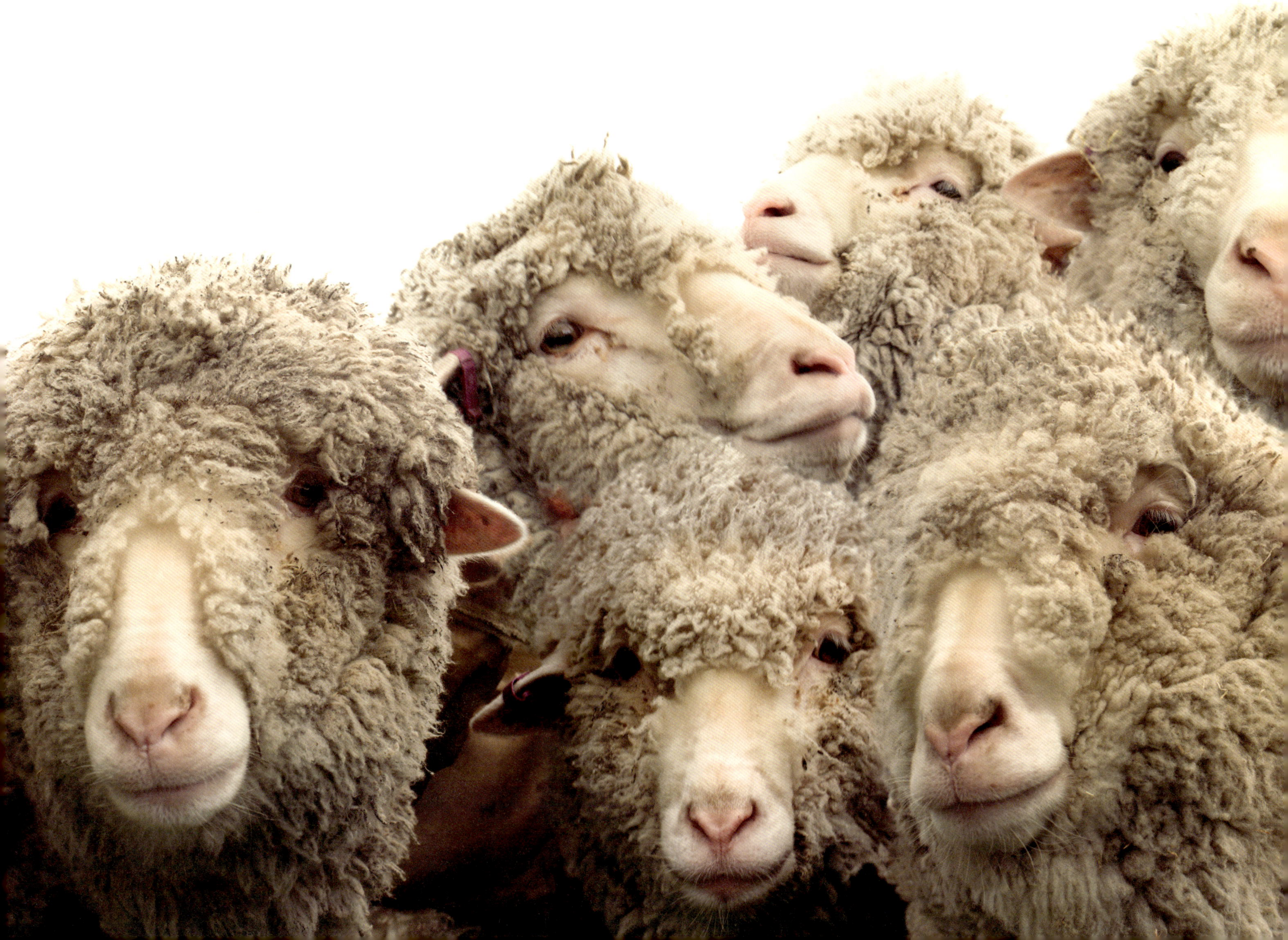

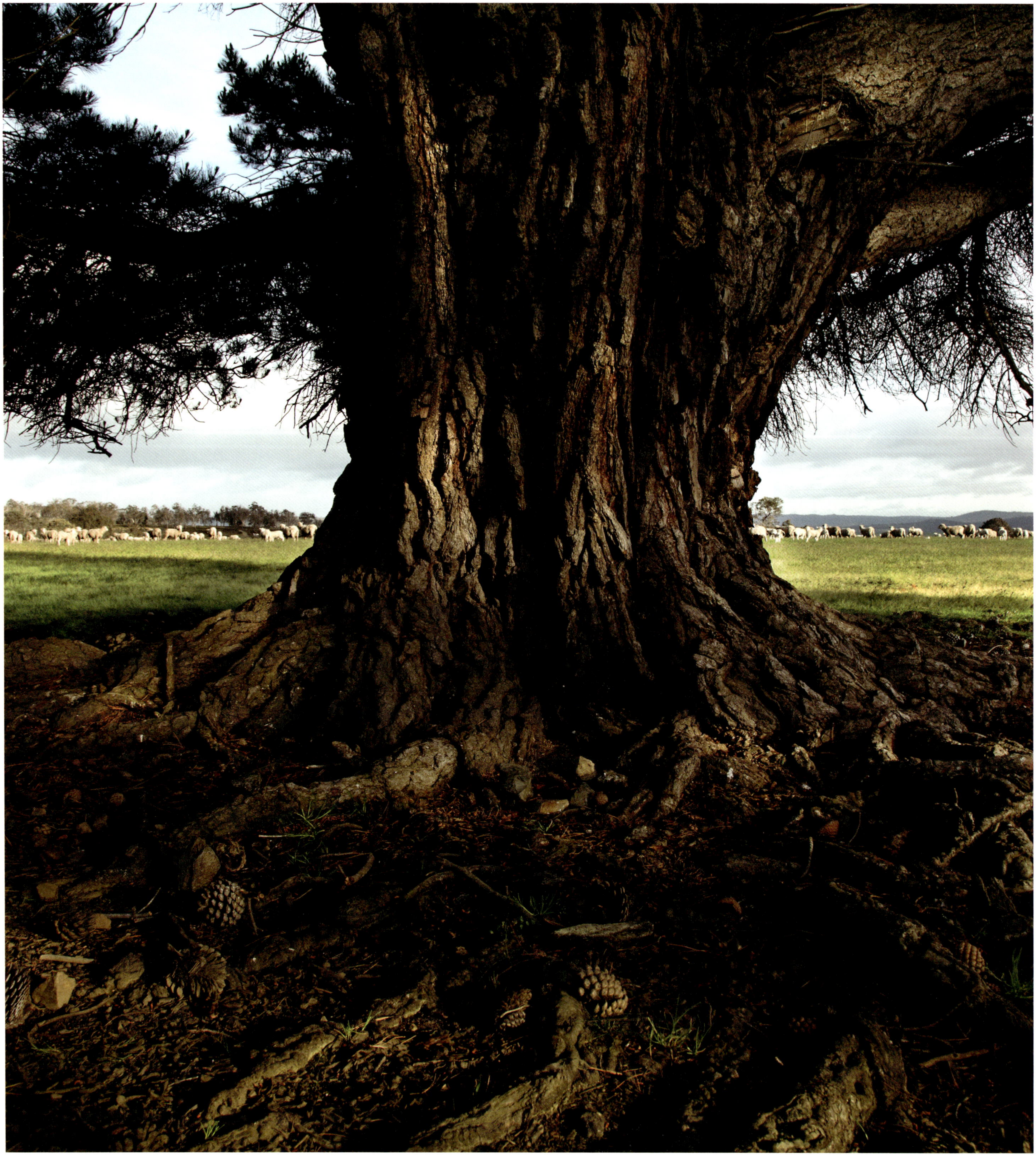

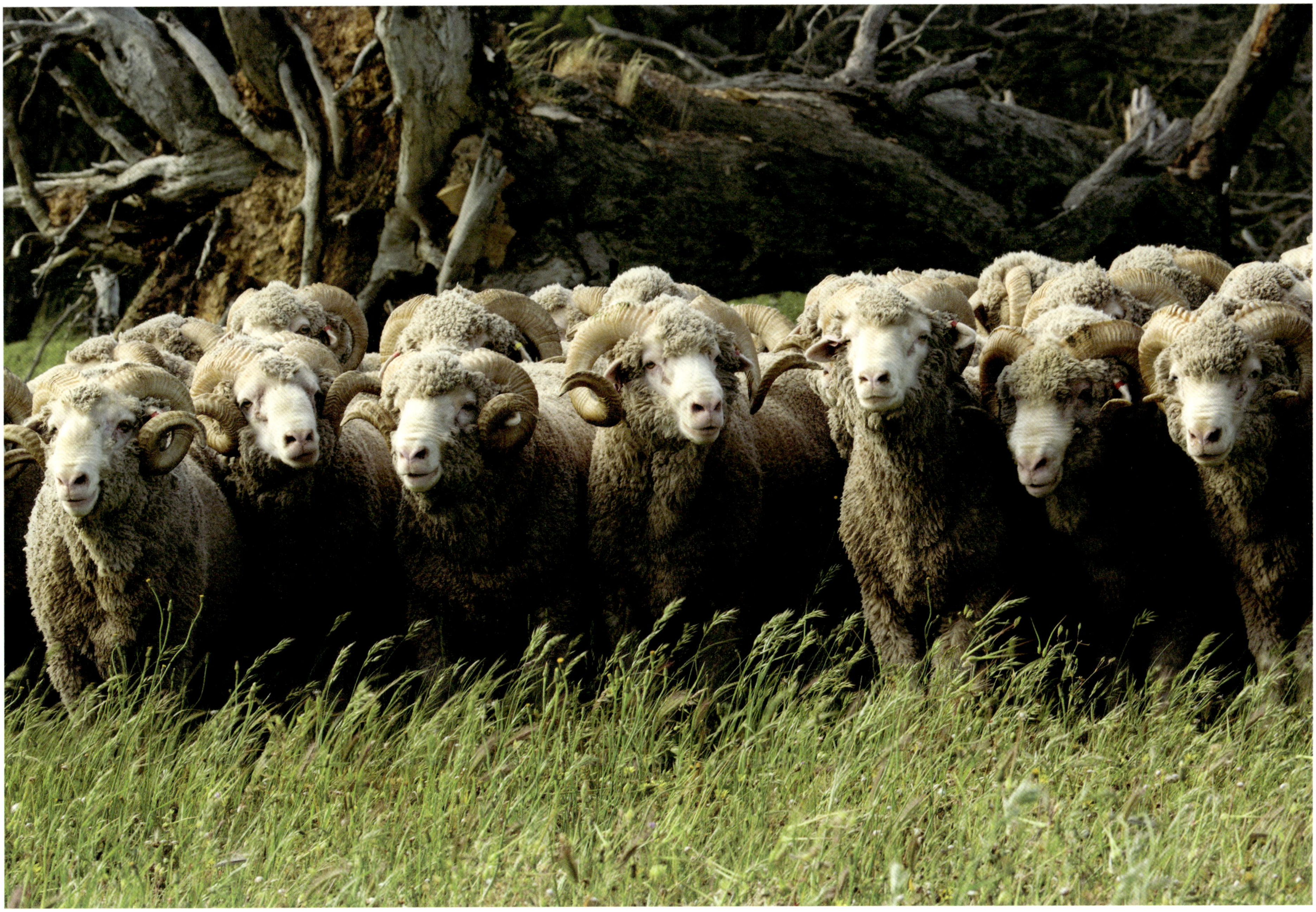

Ancient trees stand out in the vast pasturelands in Jericho, Tasmania.

So in 1979 a search got under way on a global scale for the finest bales of wool available on the market, with the aim of creating a sort of portfolio of outstanding fibre. Loro Piana travelled to Australia and New Zealand, considered the home of the merino sheep. The auctions held between September and April in the various collection points are the most common way to purchase the wool, and attract buyers from all over the world. They are the perfect place to meet the experts, genuine wool gurus, and share ideas, experiences, anecdotes and inspiration.

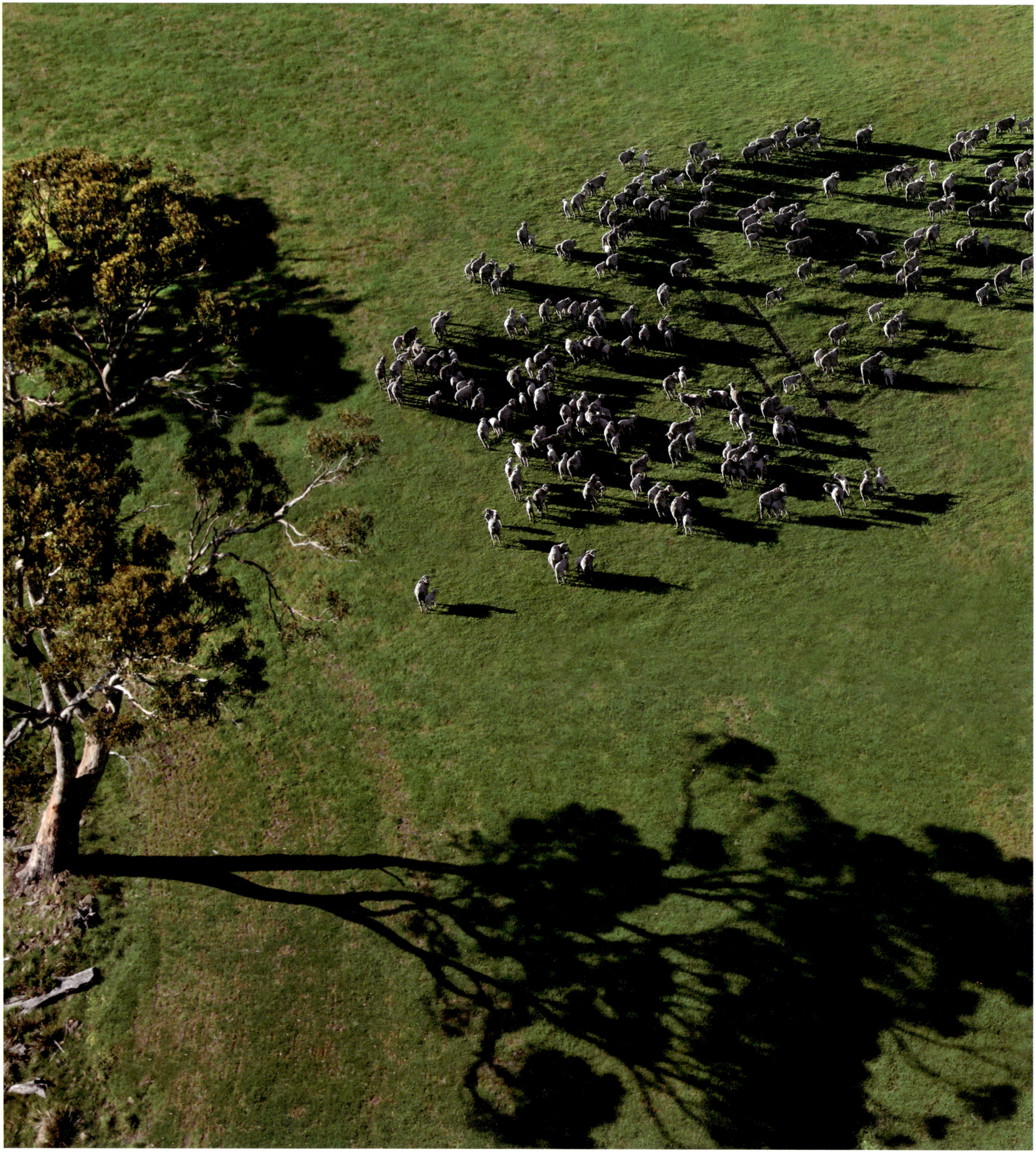

THE AREAS WHERE MERINO SHEEP ARE REARED BOAST MAJESTIC SCENERY. THE GROWERS PROTECT AND PRESERVE THESE UNSPOILT LANDS BY APPLYING THEIR EXPERT KNOWLEDGE OF THE GIFTS OF NATURE.

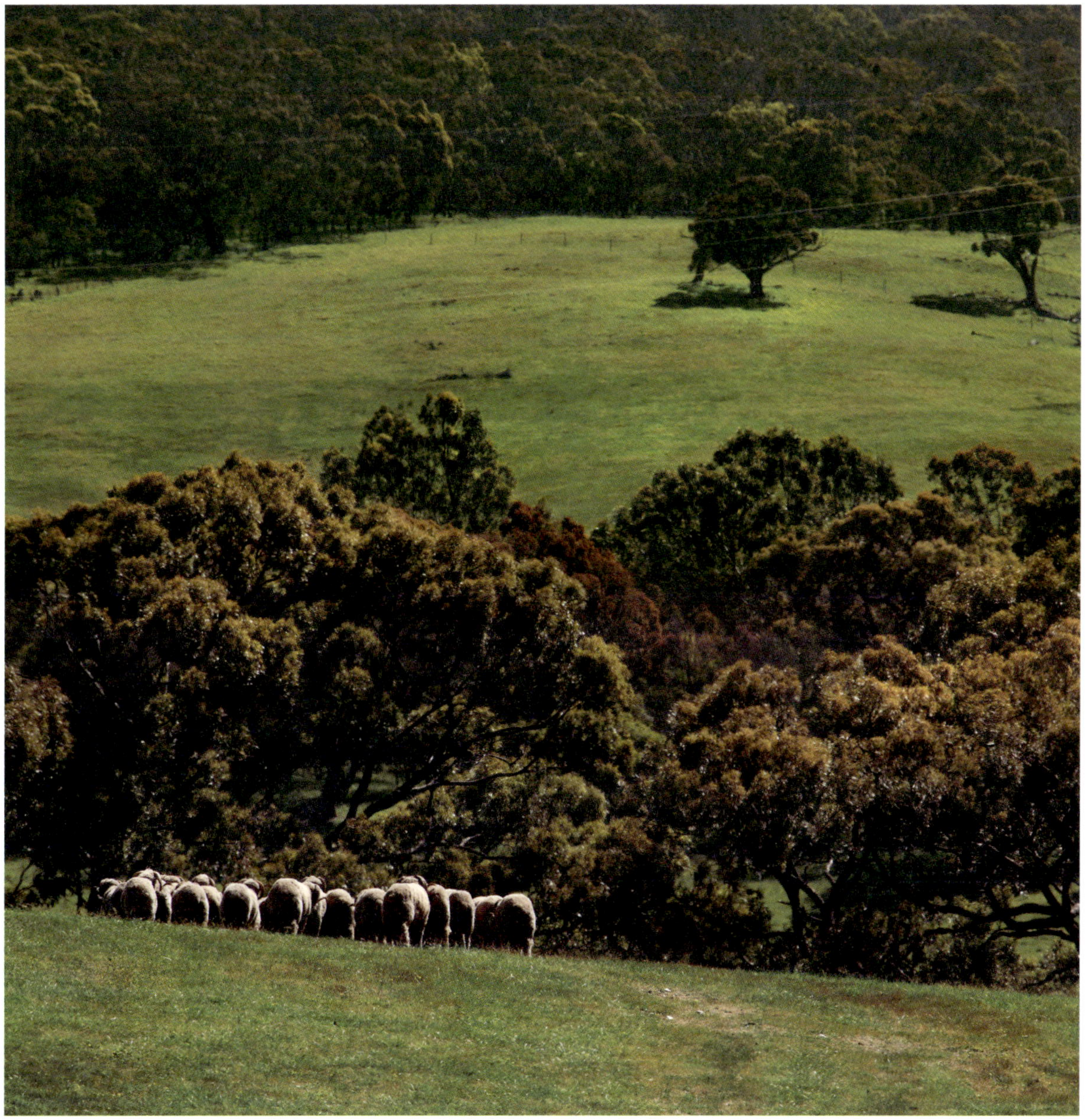

Grazing land around Oatlands, Tasmania.

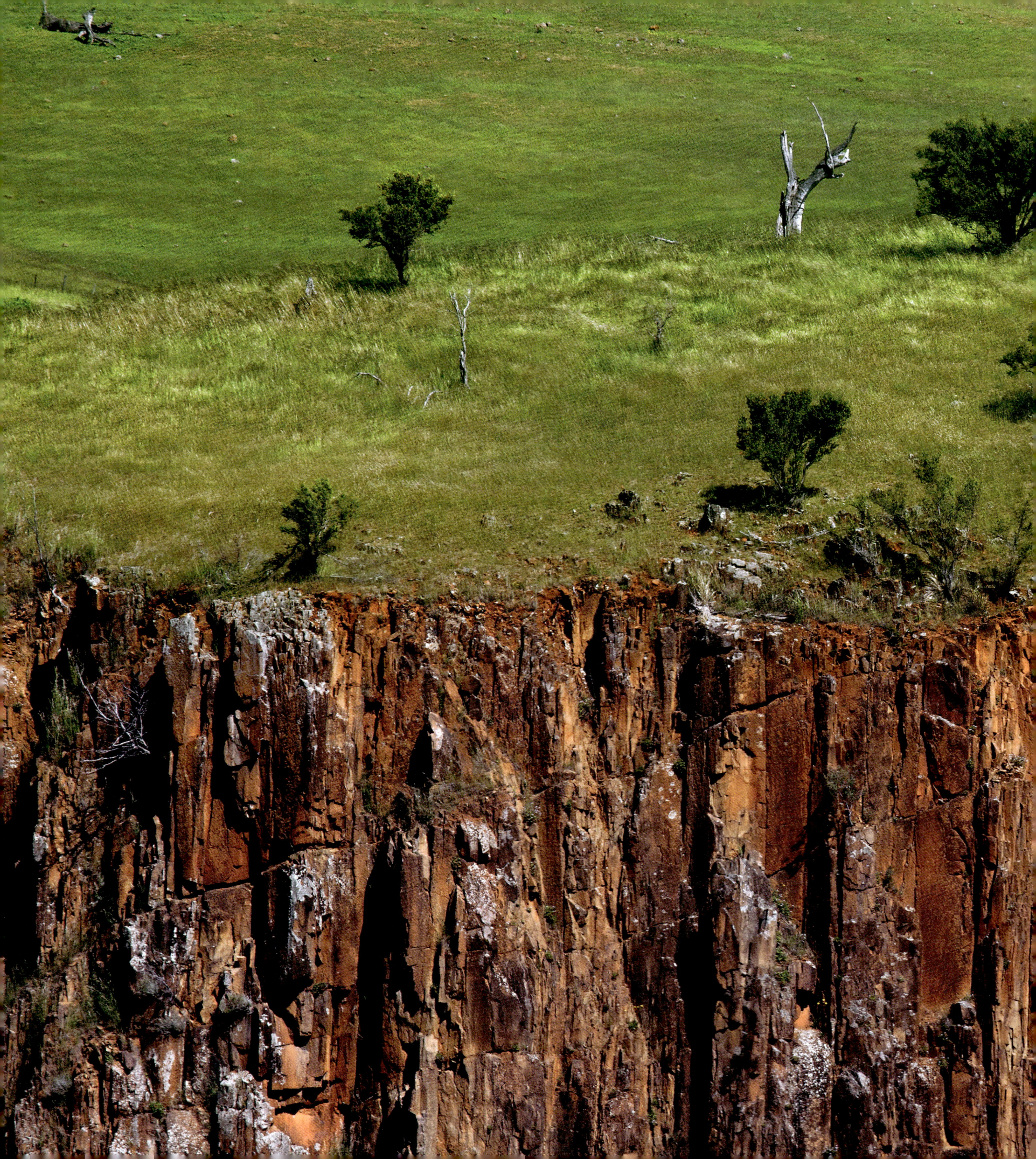

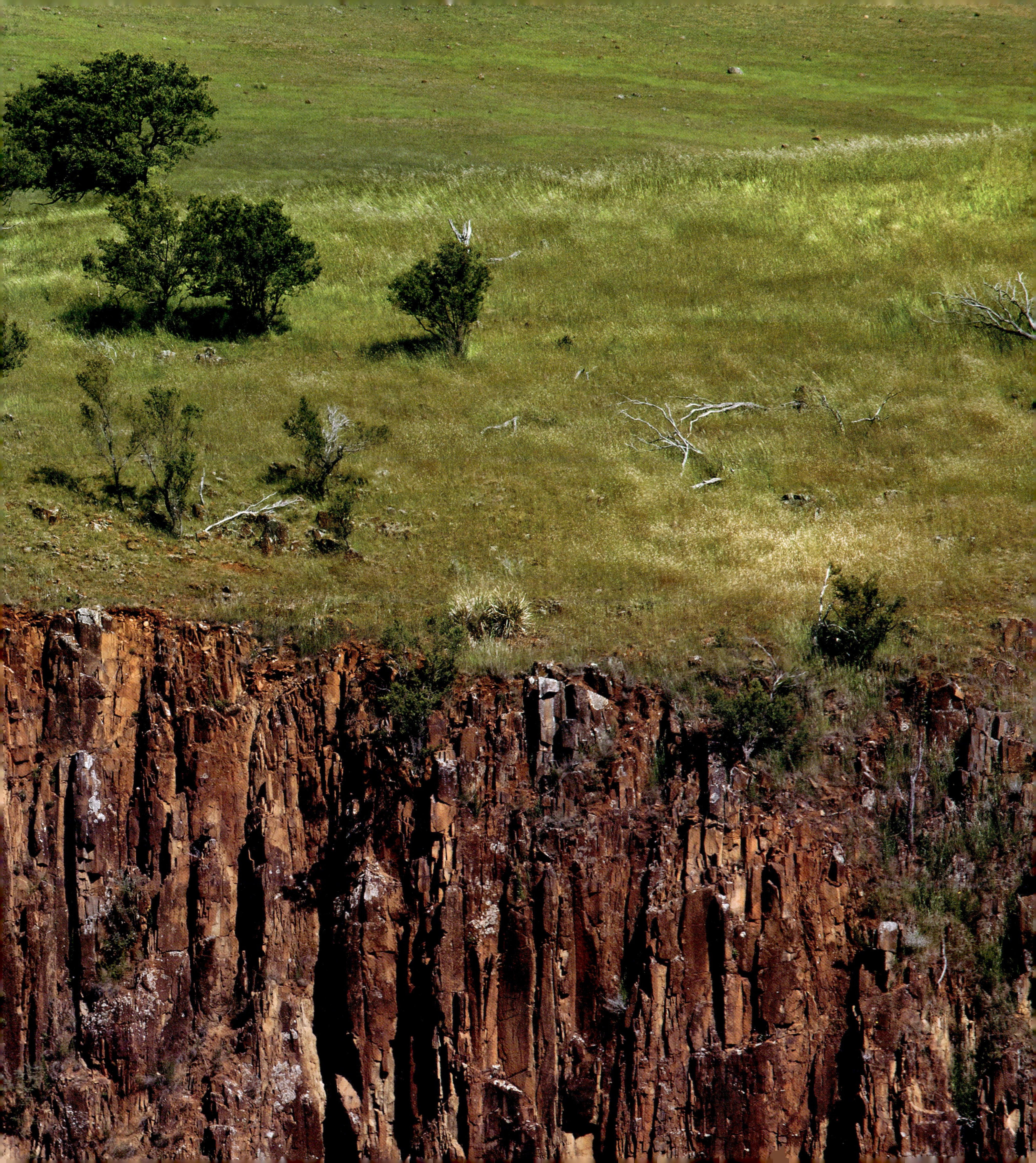

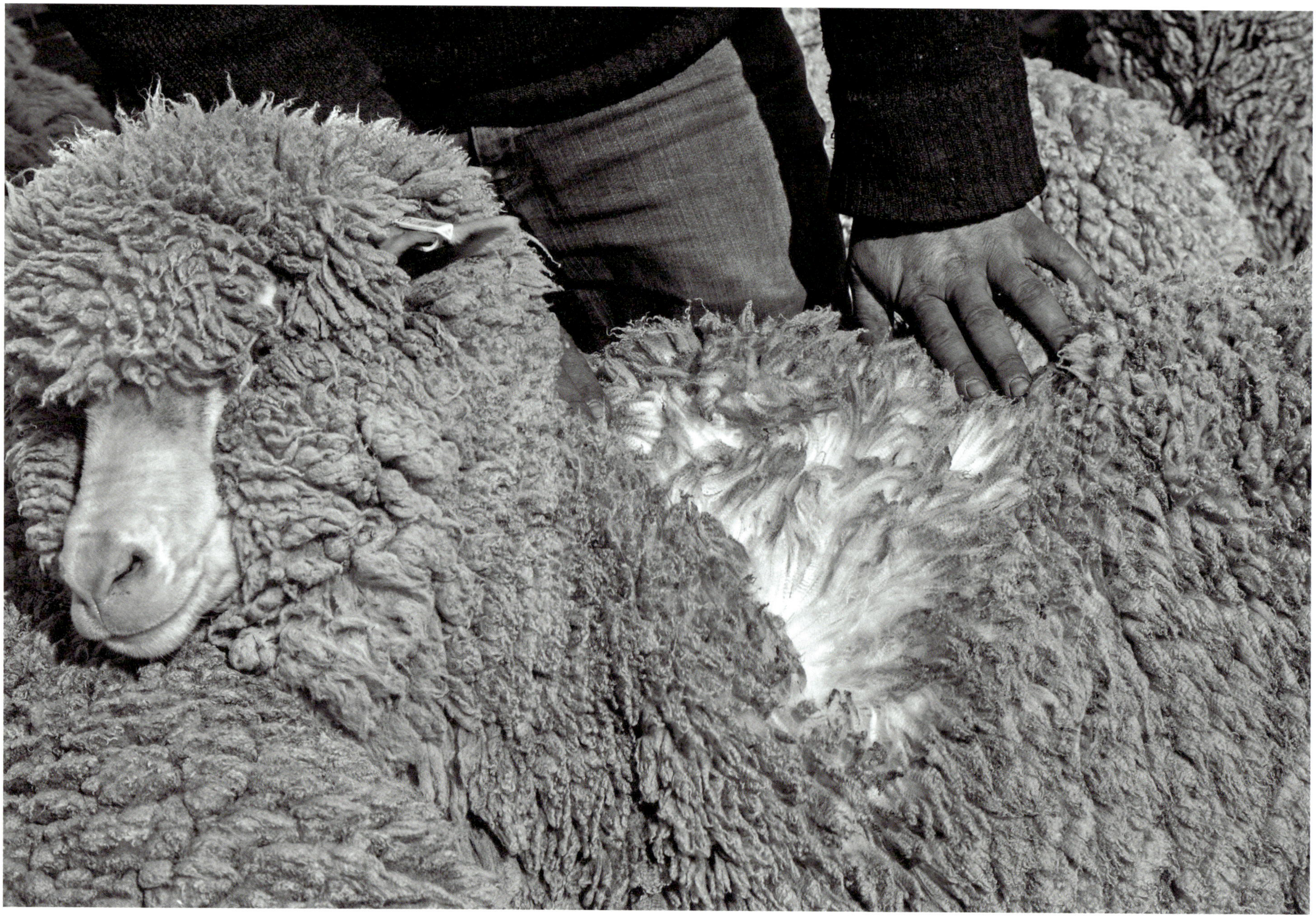

The growers who breed these merinos,
at one with nature in Tasmania,
personally take care of each animal.
Each sheep is closely monitored and
its fleece lovingly cherished.

previous pages
Towering cliffs in Jerusalem Park, Tasmania.

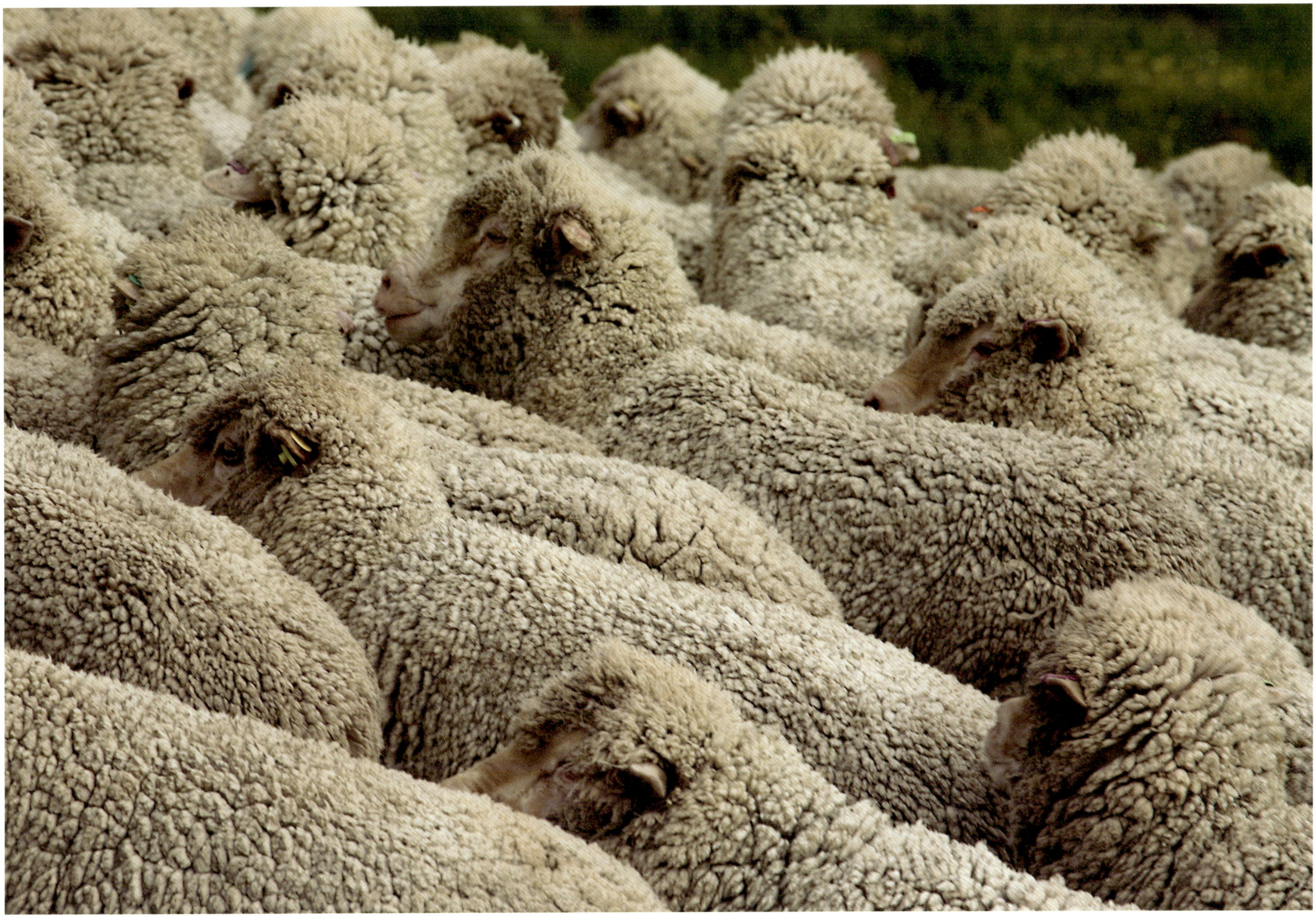

In the midst of hundreds of cubic metres of raw material of all standards were tiny quantities of a genuine treasure, a wool that reached levels of fineness that not even Loro Piana had hoped to find: no more than 30–50 special bales out of approximately 4 million up for auction each year, classified Top Line 1PP (1PP = 1 Plus Plus, the very highest level of fineness and style).

Franco Loro Piana's warning about the potential extinction of extrafine wool came to mind. The very existence of those few bales of supreme wool proved that there were farmers who, like Loro Piana, were pursuing an uncompromising standard of quality. It was time to find them.

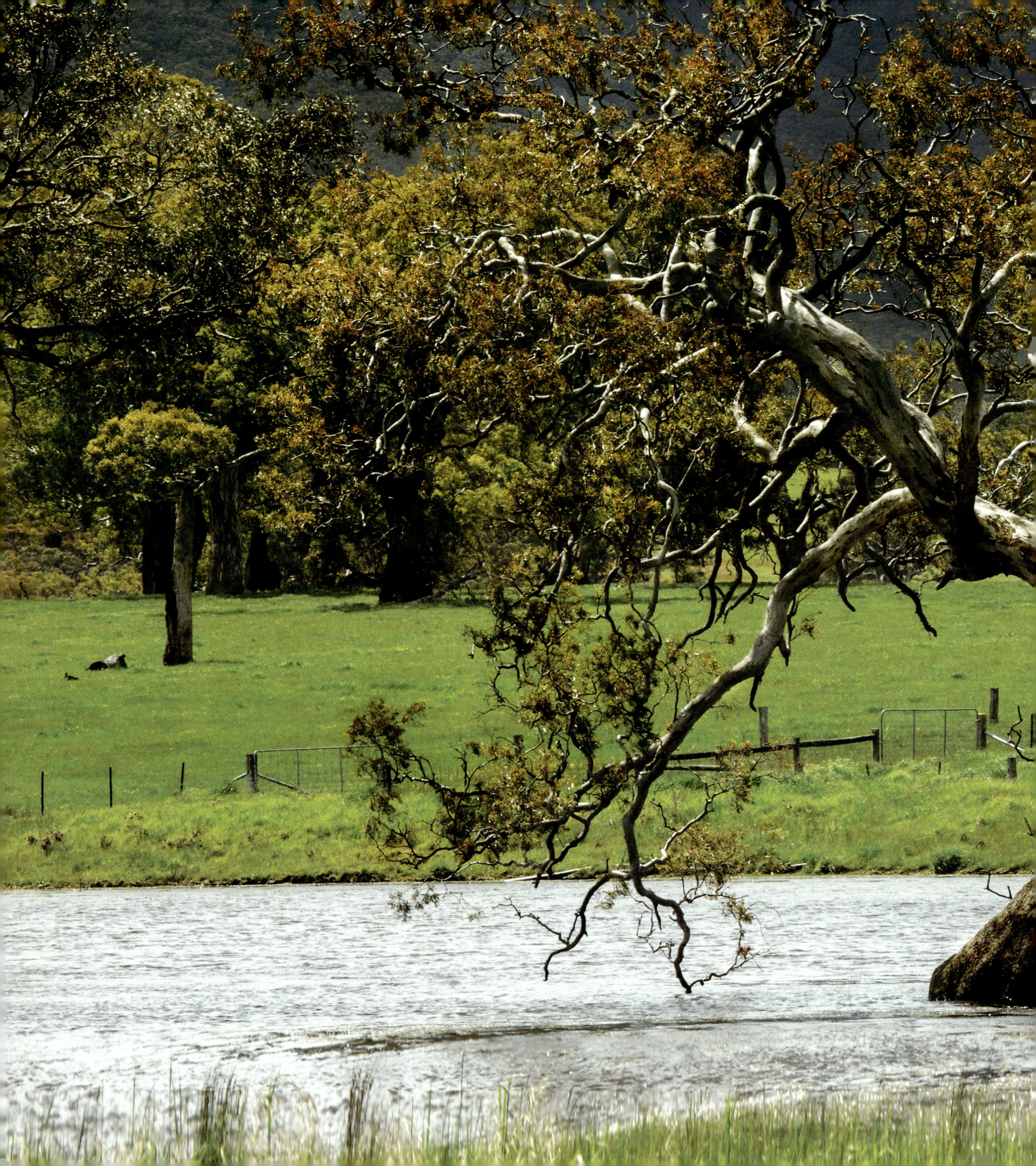

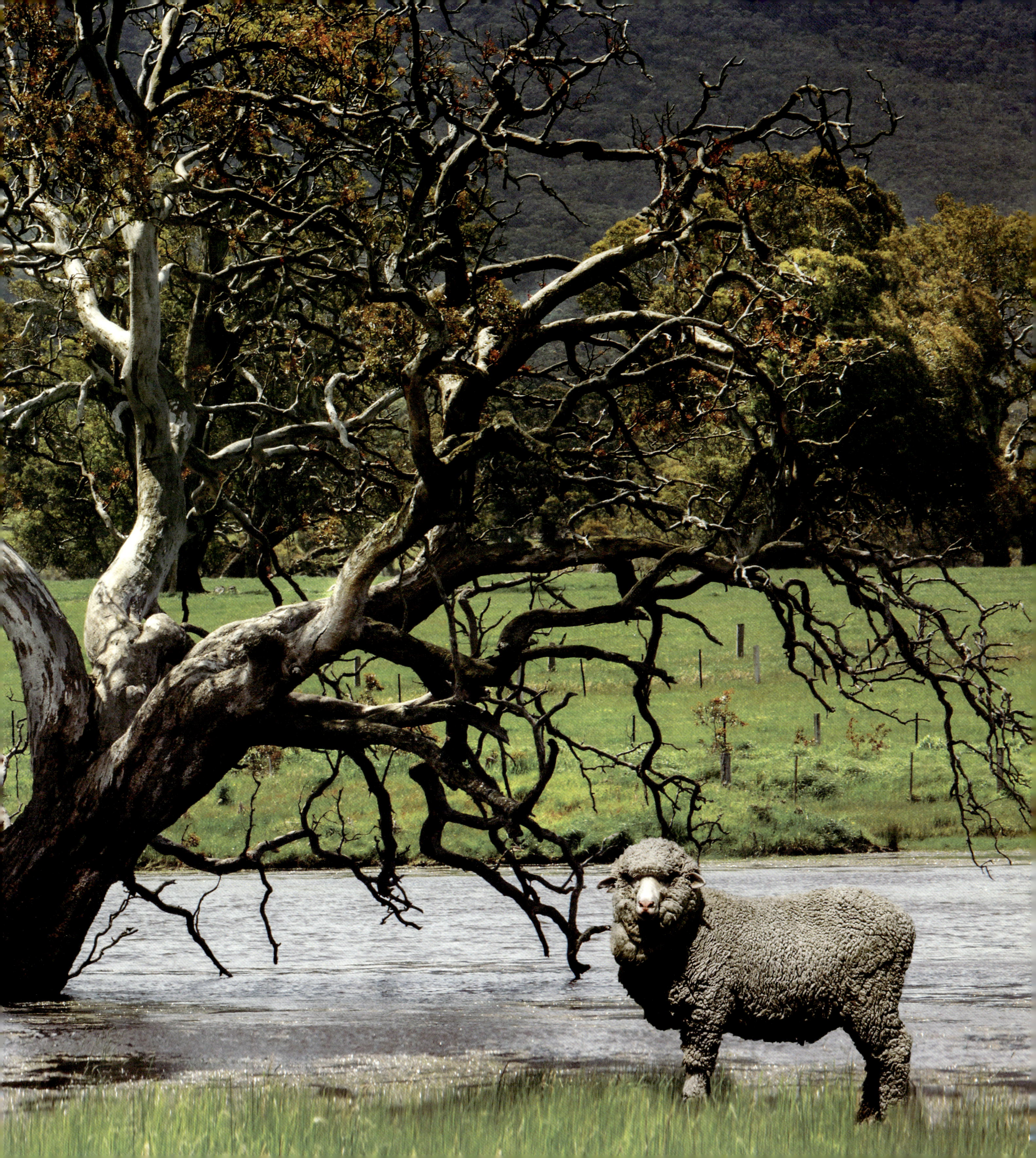

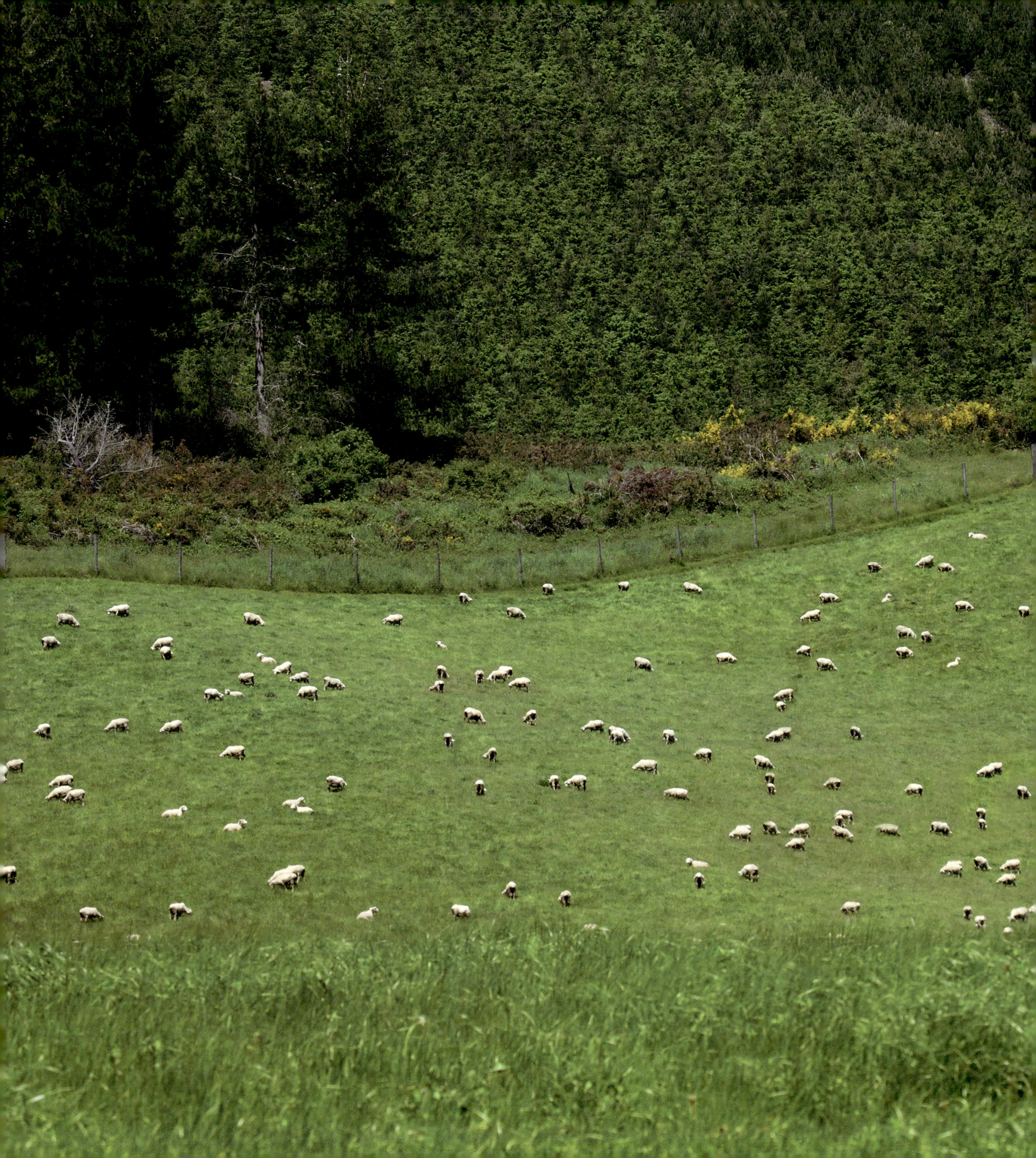

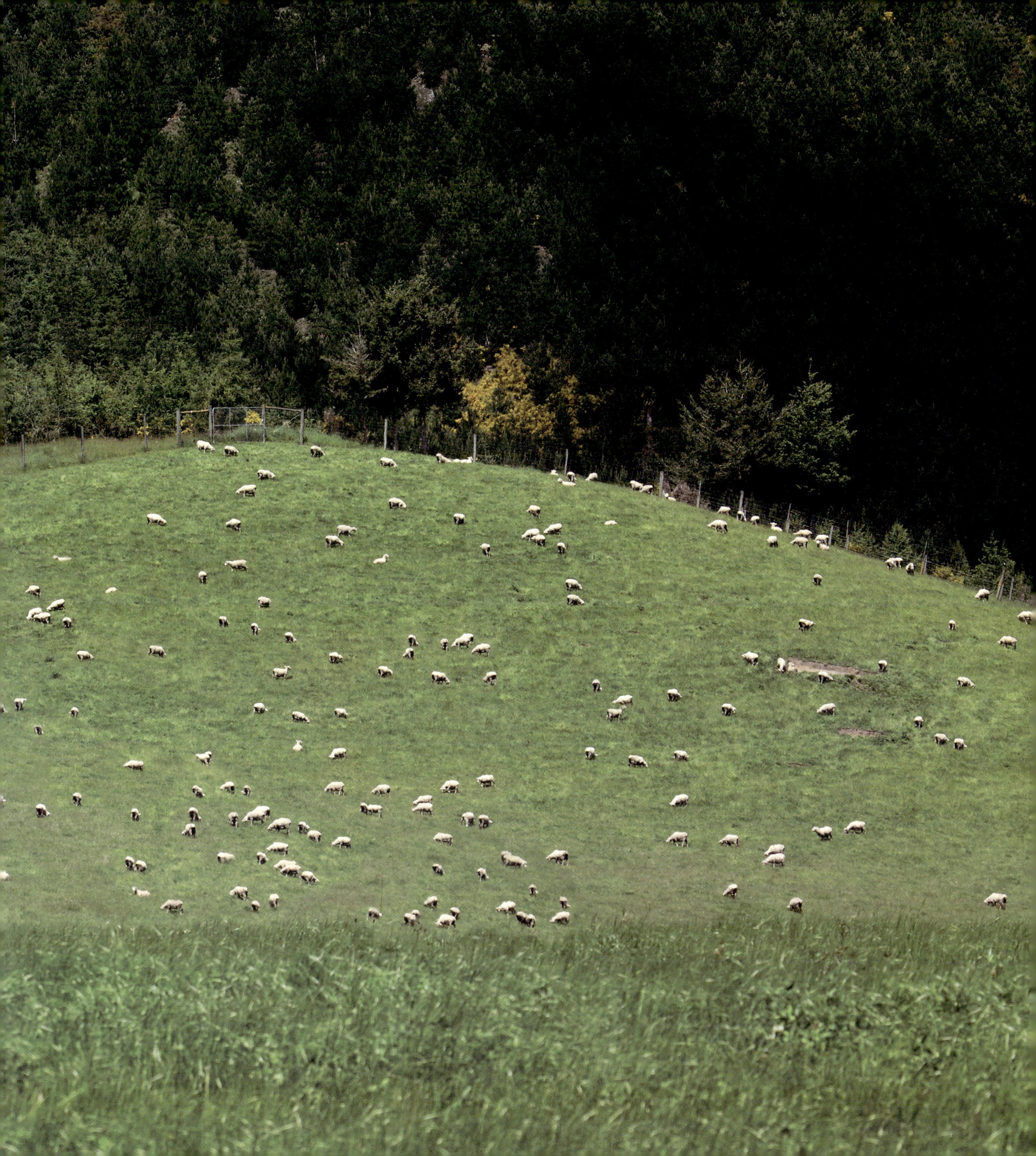

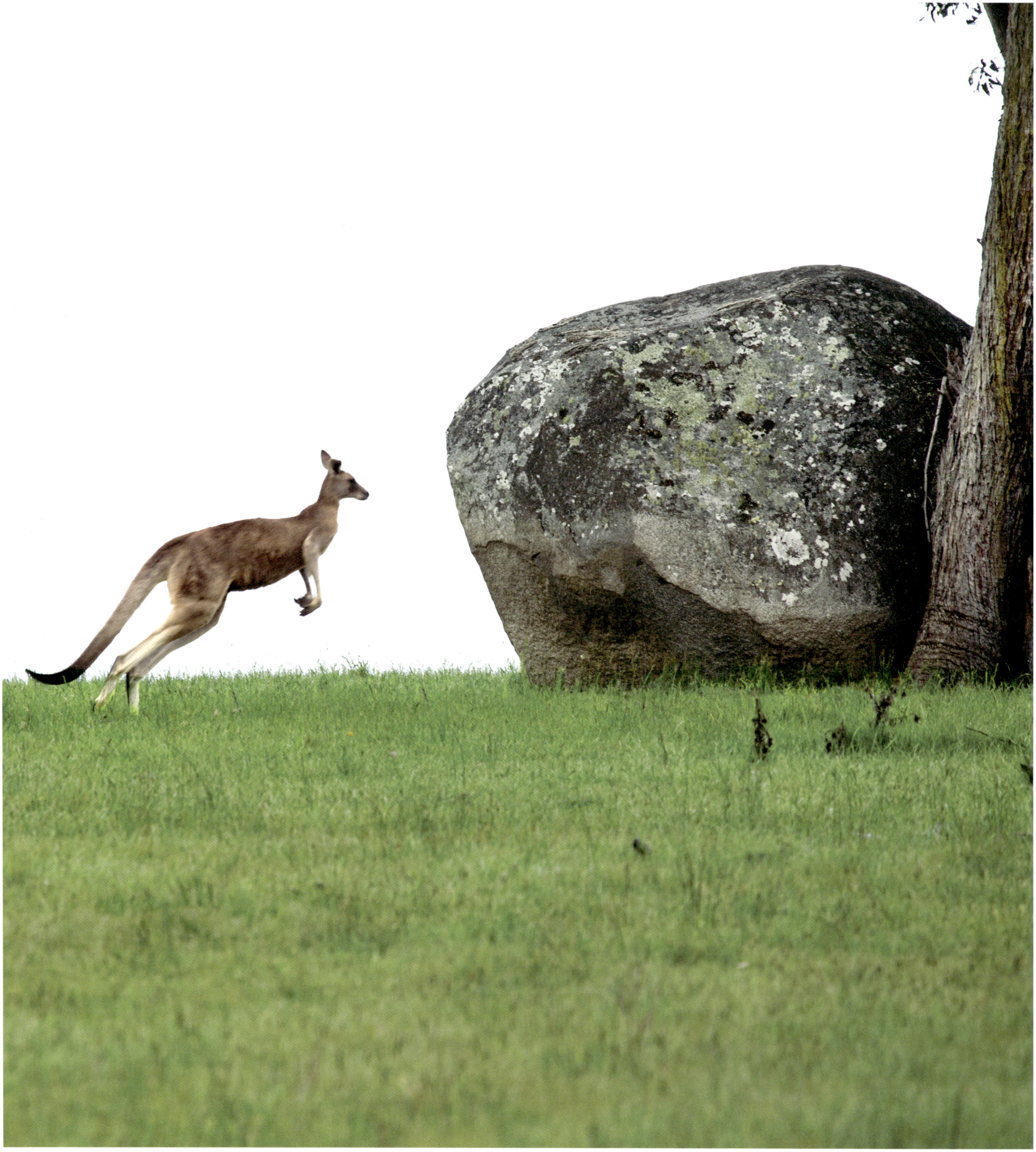

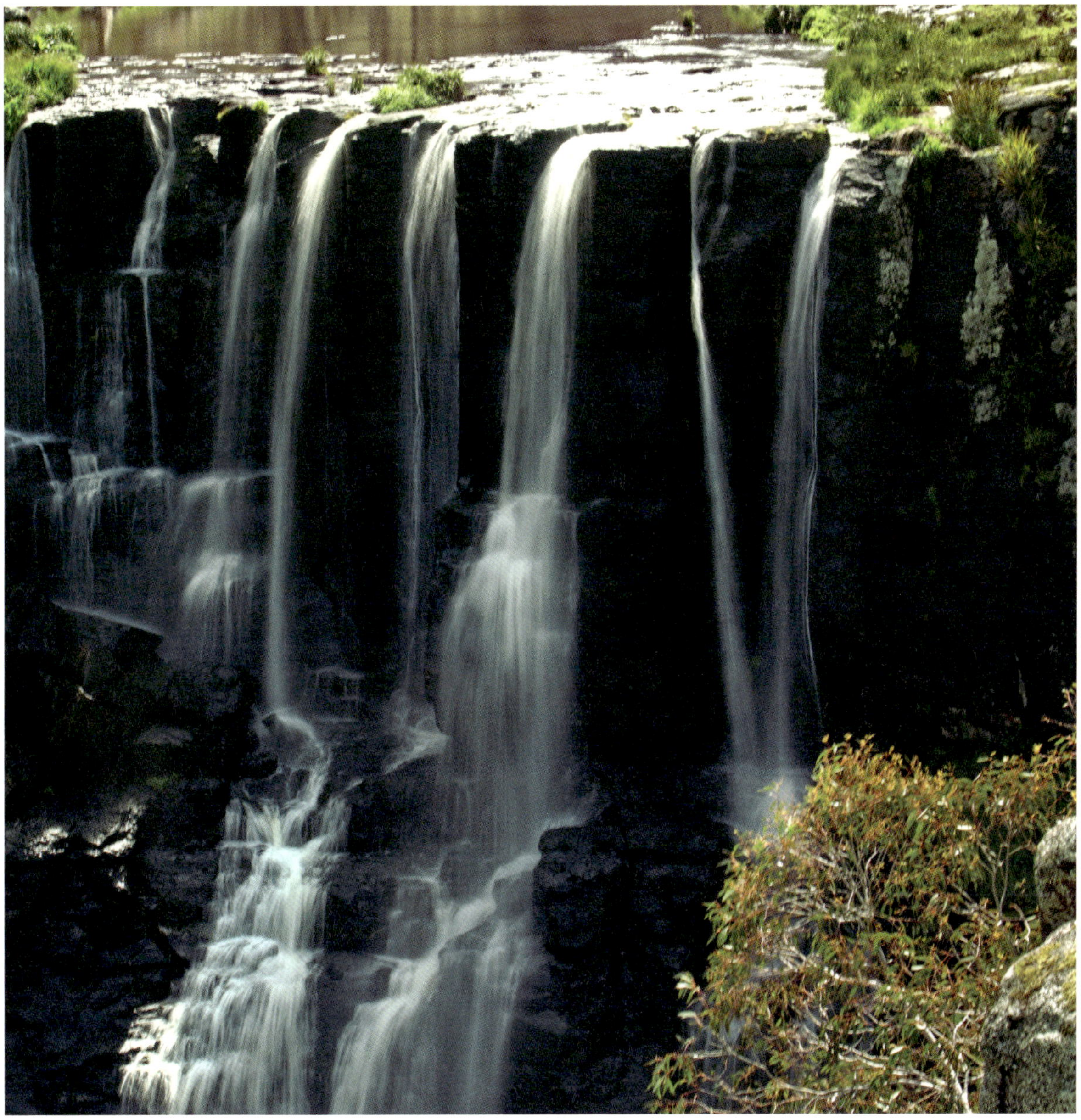

AUSTRALIA'S WATERFALLS ARE A TRULY BREATHTAKING SIGHT, AND THE PRIDE AND JOY OF THE LOCAL PEOPLE. THE PHOTOGRAPH SHOWS EBOR FALLS IN NEW SOUTH WALES.

Kangaroos, the national symbol of Australia along with sheep, move among the ancient, towering rocks of the Cathedral Rock National Park.

following pages
Grazing land near Armidale, Australia.

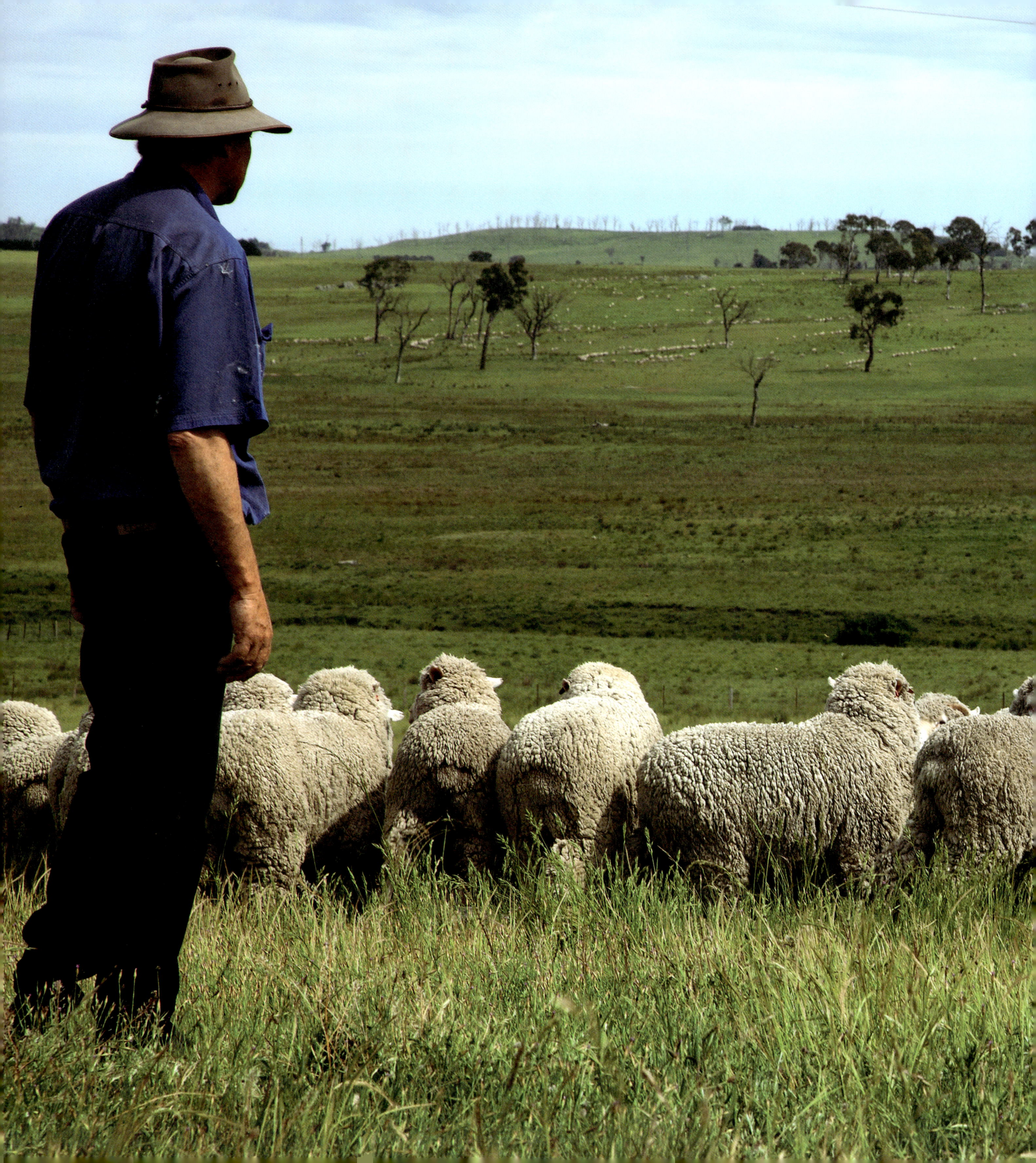

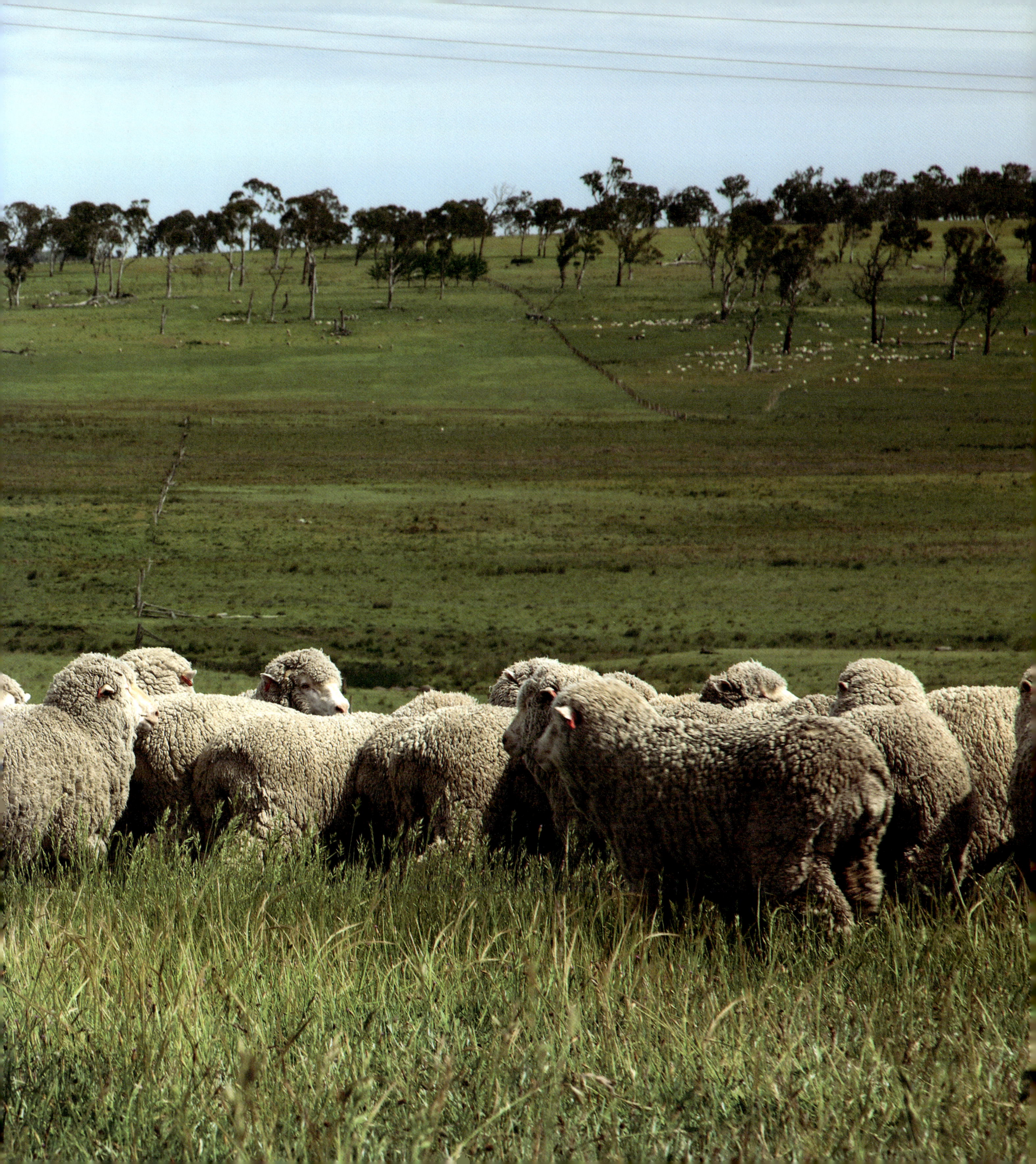

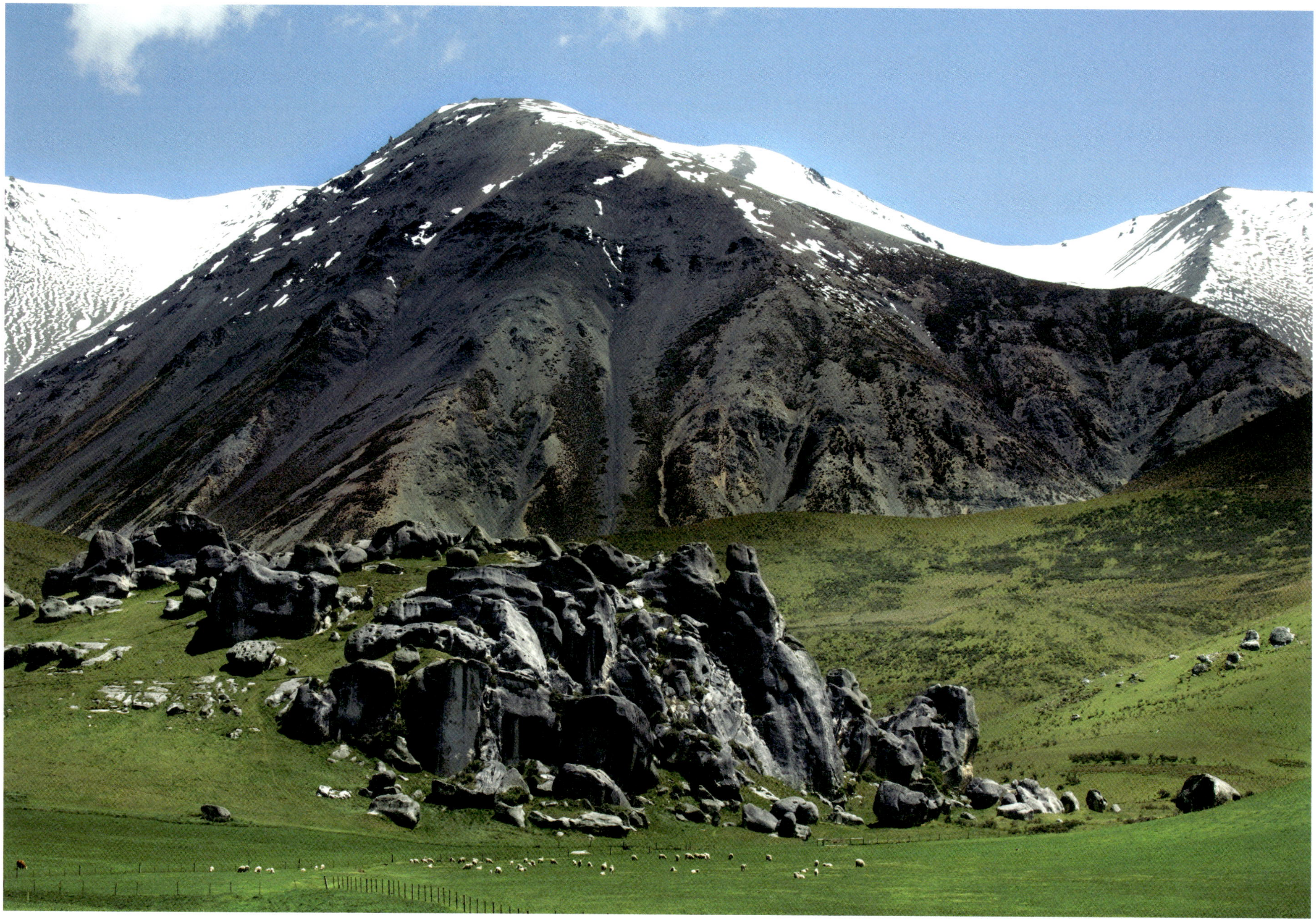

Grazing land around Castle Rock, New Zealand
and Dunkeld, Australia.

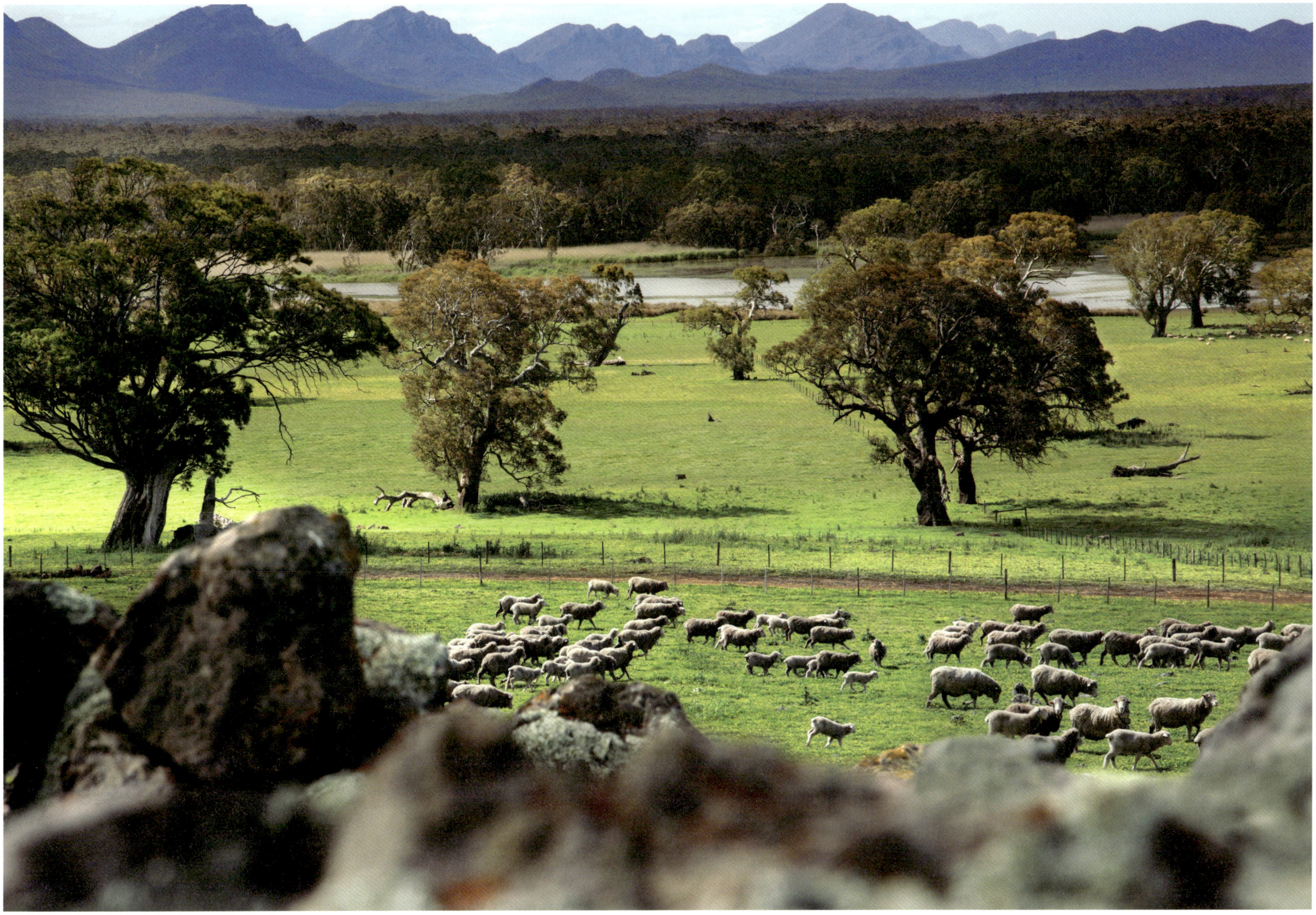

Loro Piana set about contacting the growers directly and visiting them at home, on their farms: magical places, set in unspoilt countryside, among rivers and waterfalls, forests and rocky sea cliffs. This is the land that for centuries has been home to the world's oldest sheep breed, the merino. In these places merino sheep are a genuine asset, and seem to have been part of the landscape forever, their dense, crimped fleeces blending into the background of rocks, ancient mountainsides and cliffs.

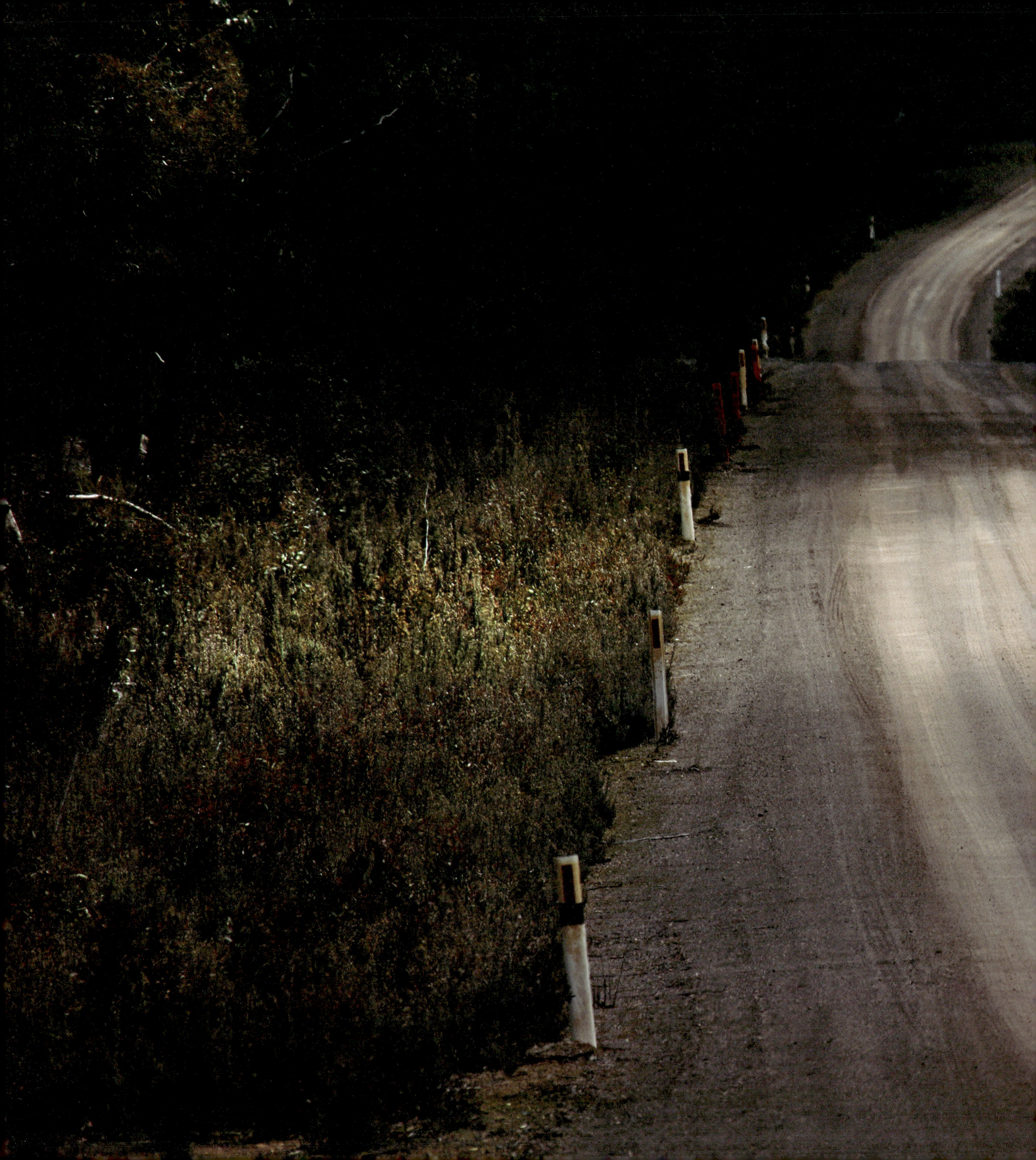

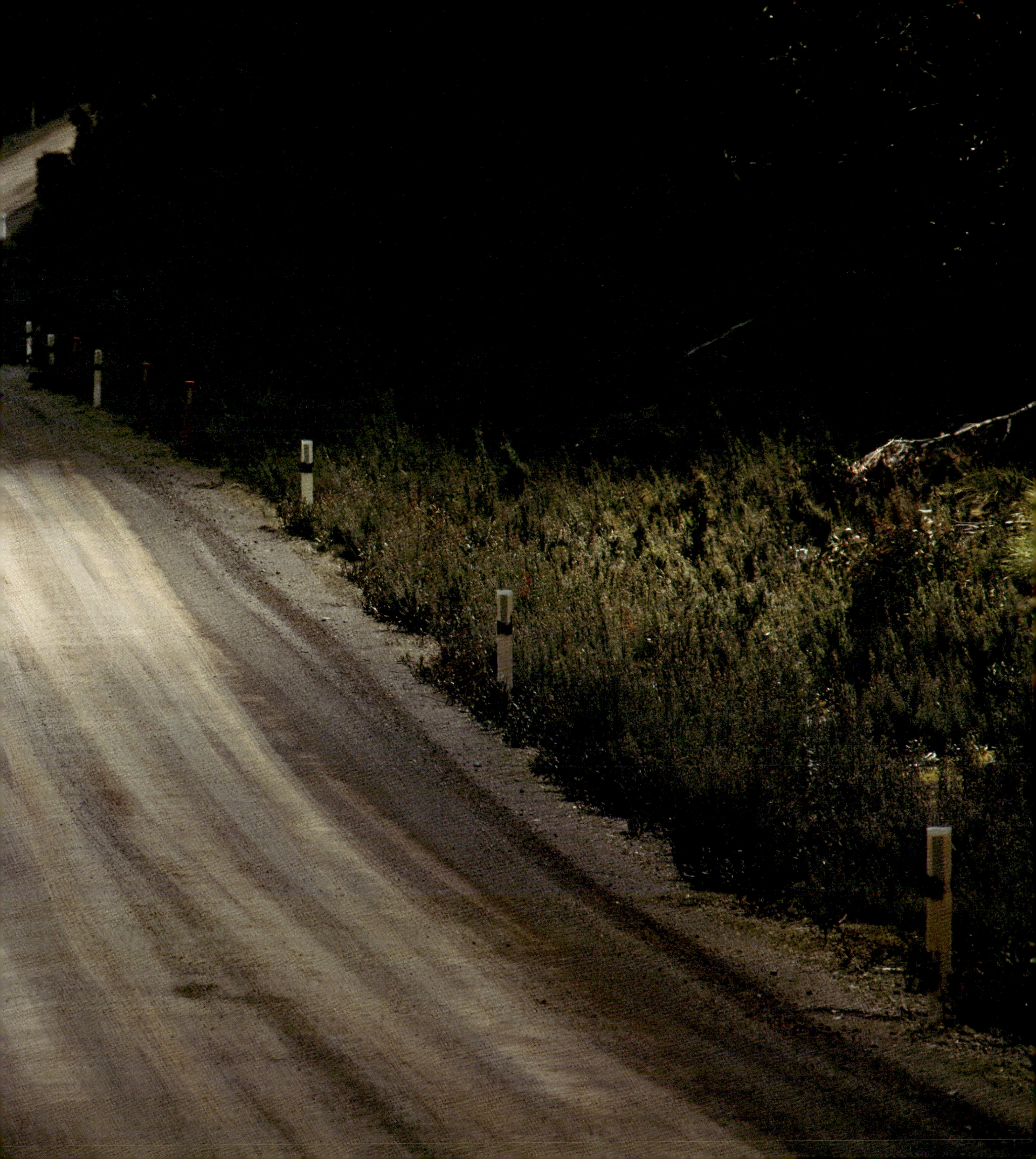

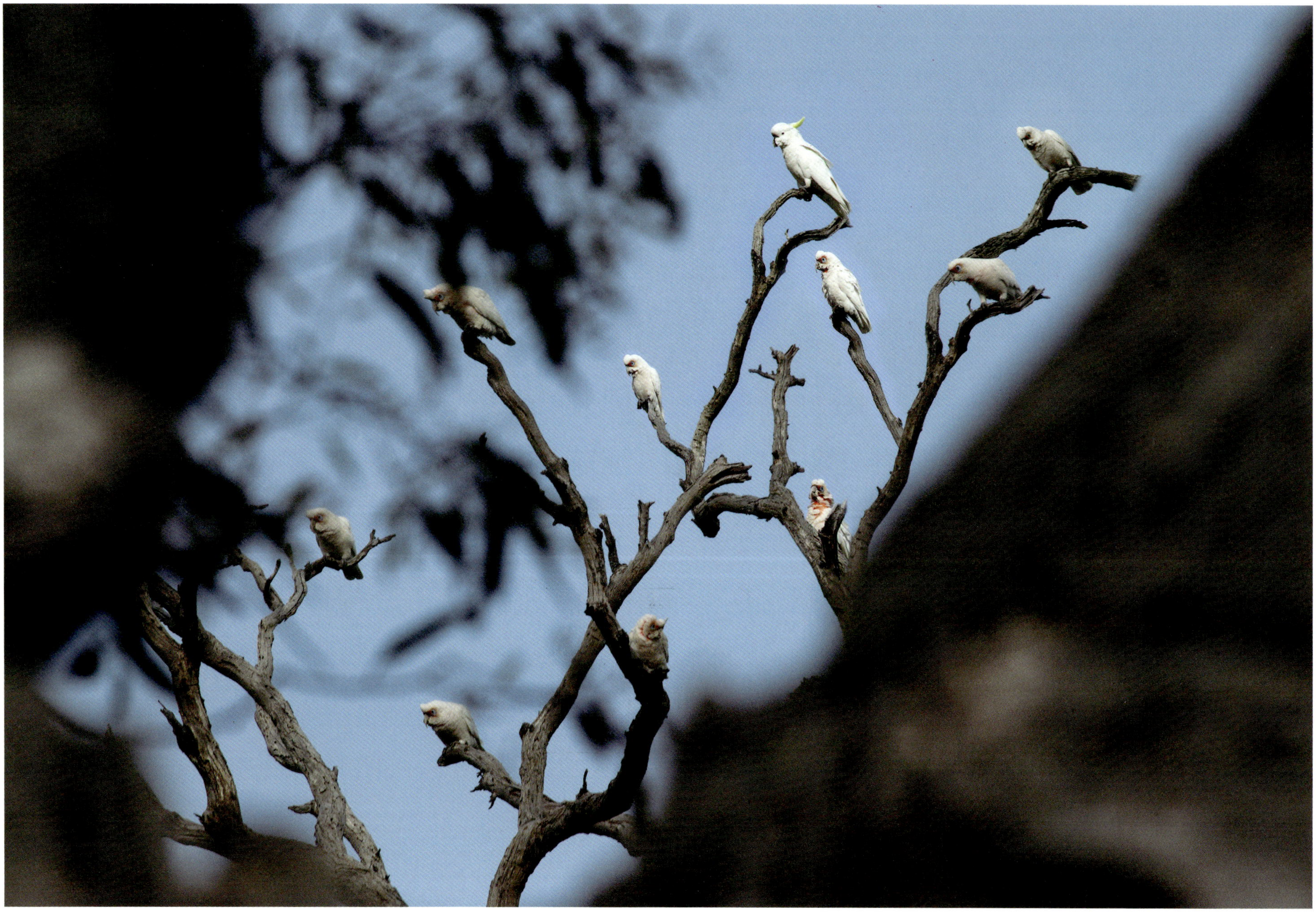

Enfield State Park, Australia.

previous pages
These endless roads hint at mysteries,
discoveries and adventure
in the rugged beauty of Tasmania.

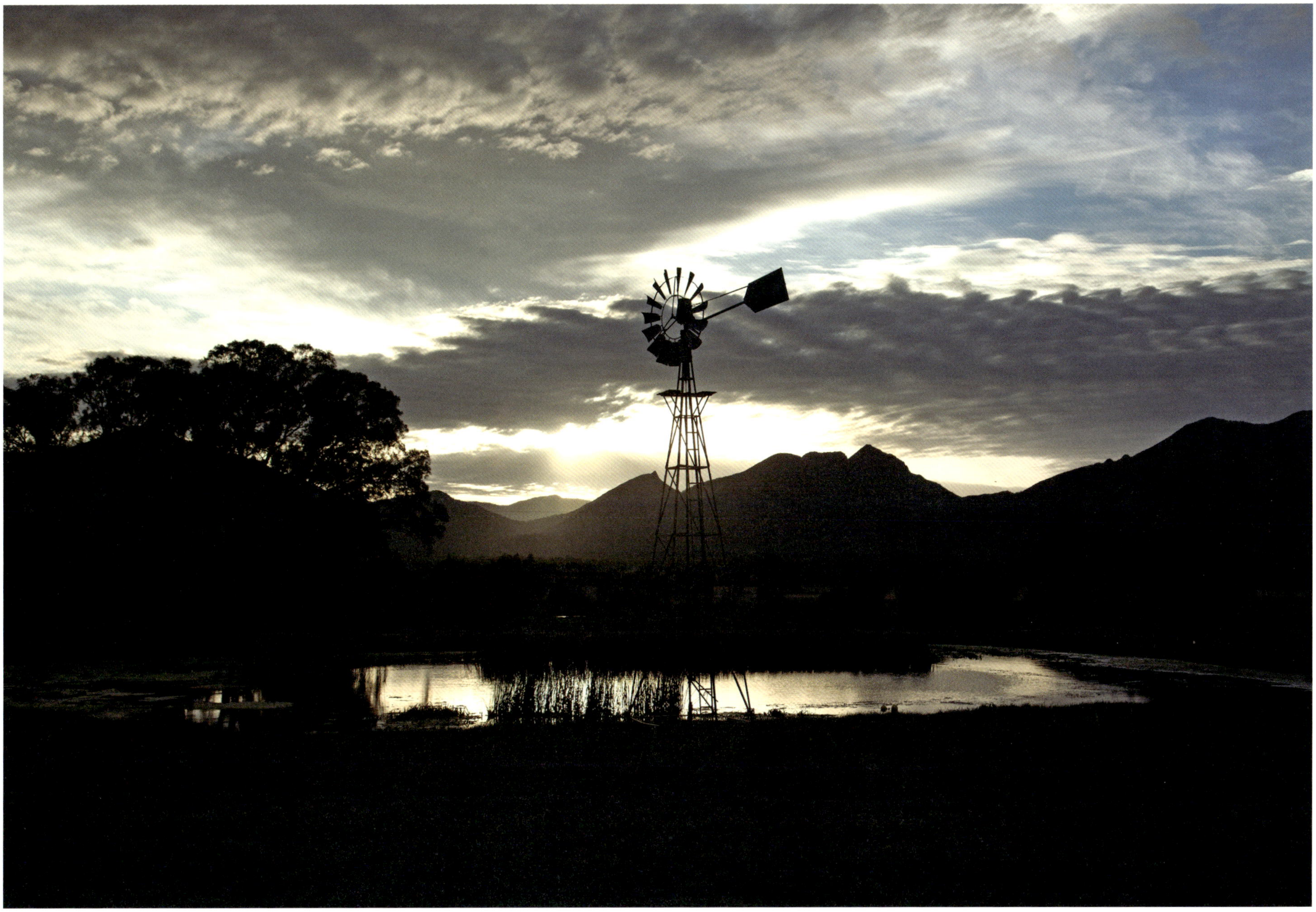

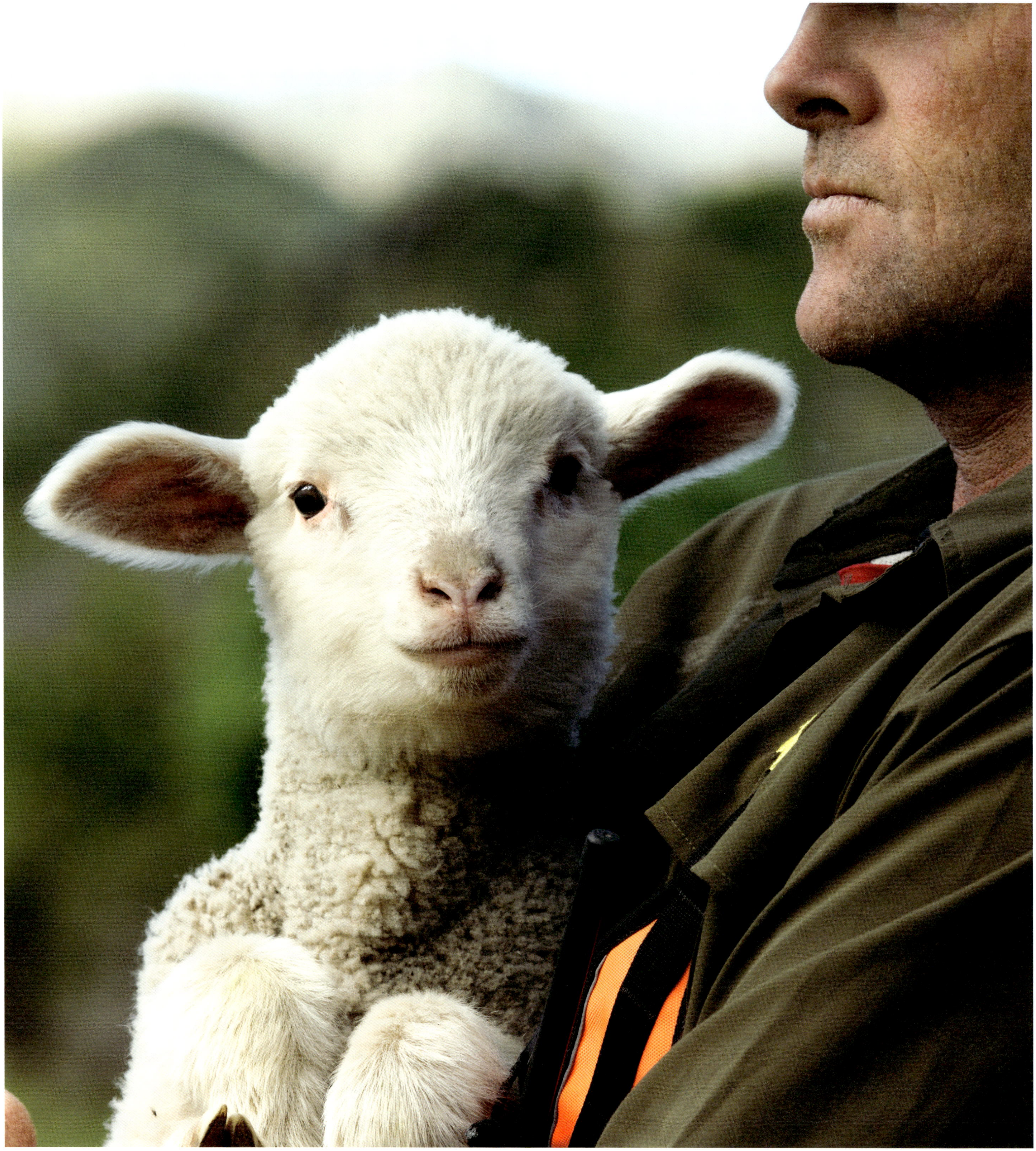

WHETHER IN ARMIDALE,

NEW SOUTH WALES, OR TASMANIA,

ON THE FARMS EVERYONE

IS PERSONALLY INVOLVED

IN LOOKING AFTER THE SHEEP

AND THE WOOL THEY PRODUCE.

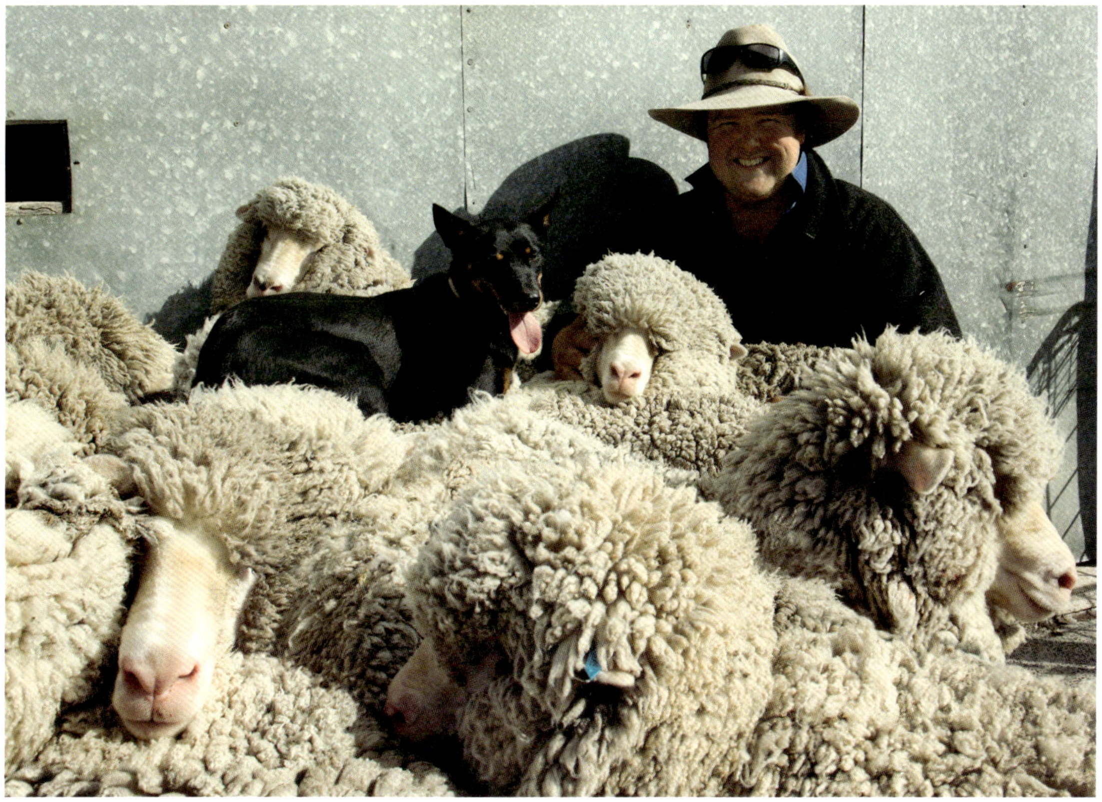

On visiting the farms, Loro Piana buyers, who were used to purchasing their wool at the auctions, were greeted with a sight out of time. The growers opened their houses and proudly showed their lands and their cherished animals, treated like part of the family, each sheep with a name, a history, a personality. Everything had a familiar air and there was the distinct impression that in these parts work was more than just a job, it was a genuine way of life involving the whole community.

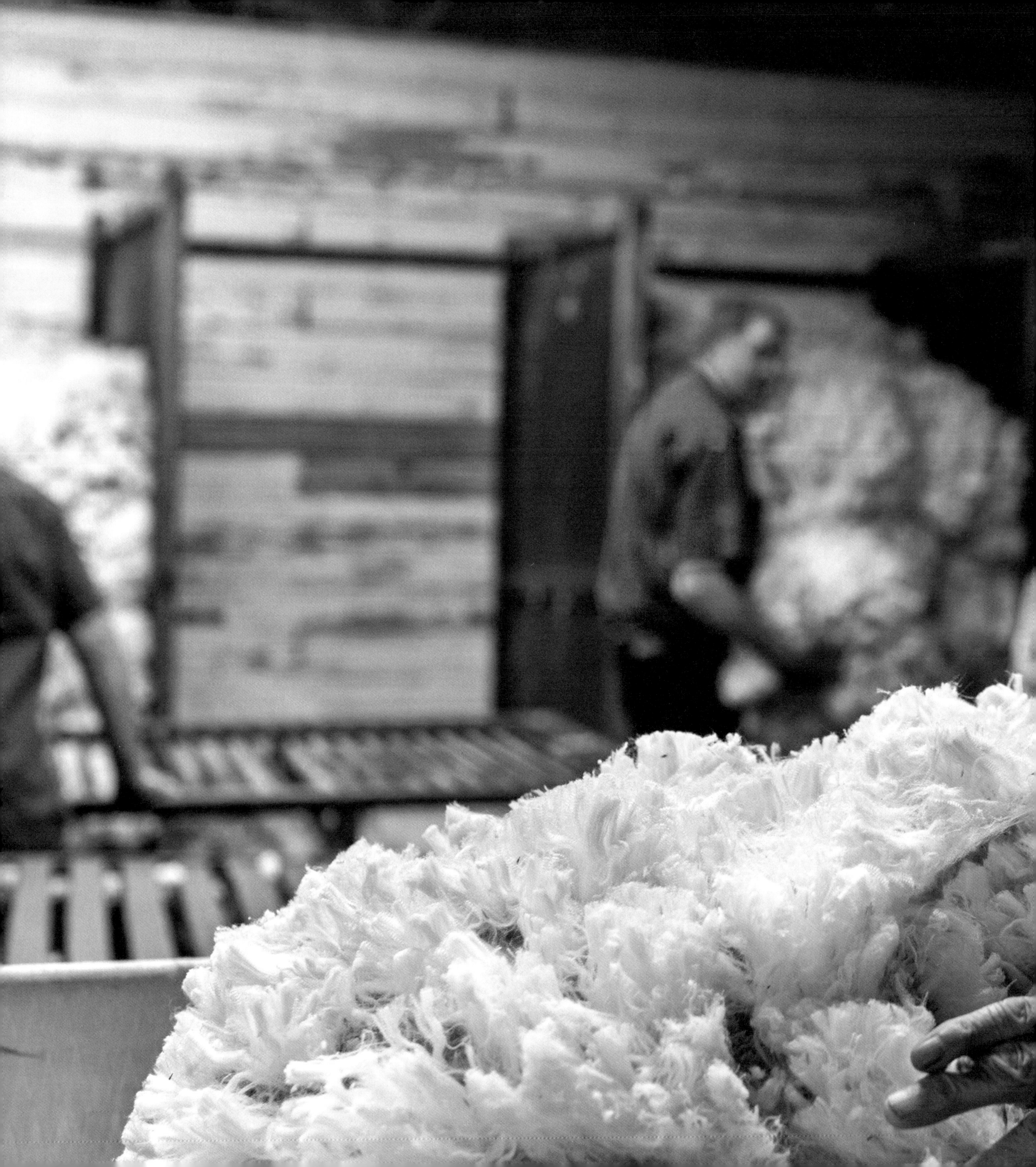

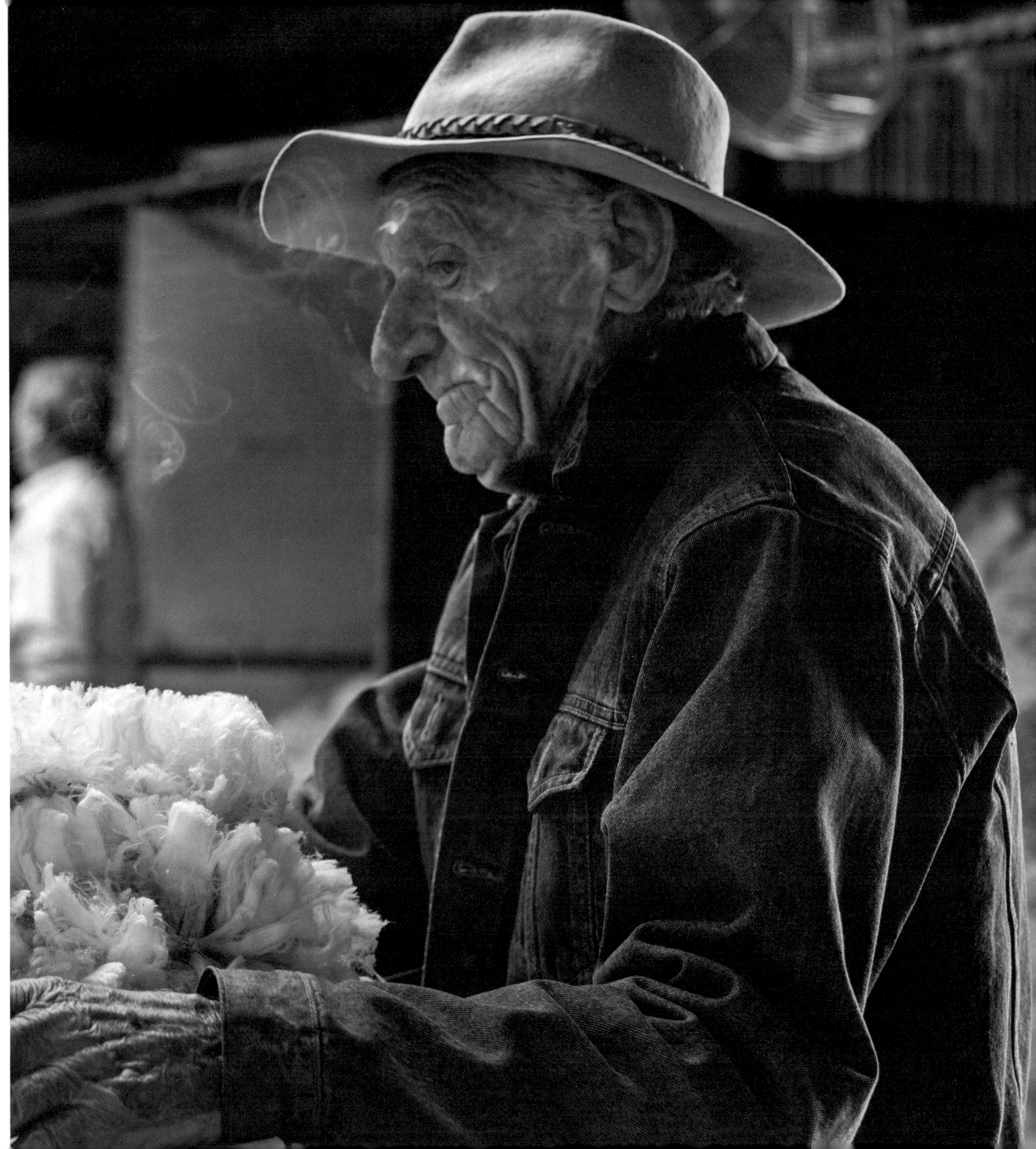

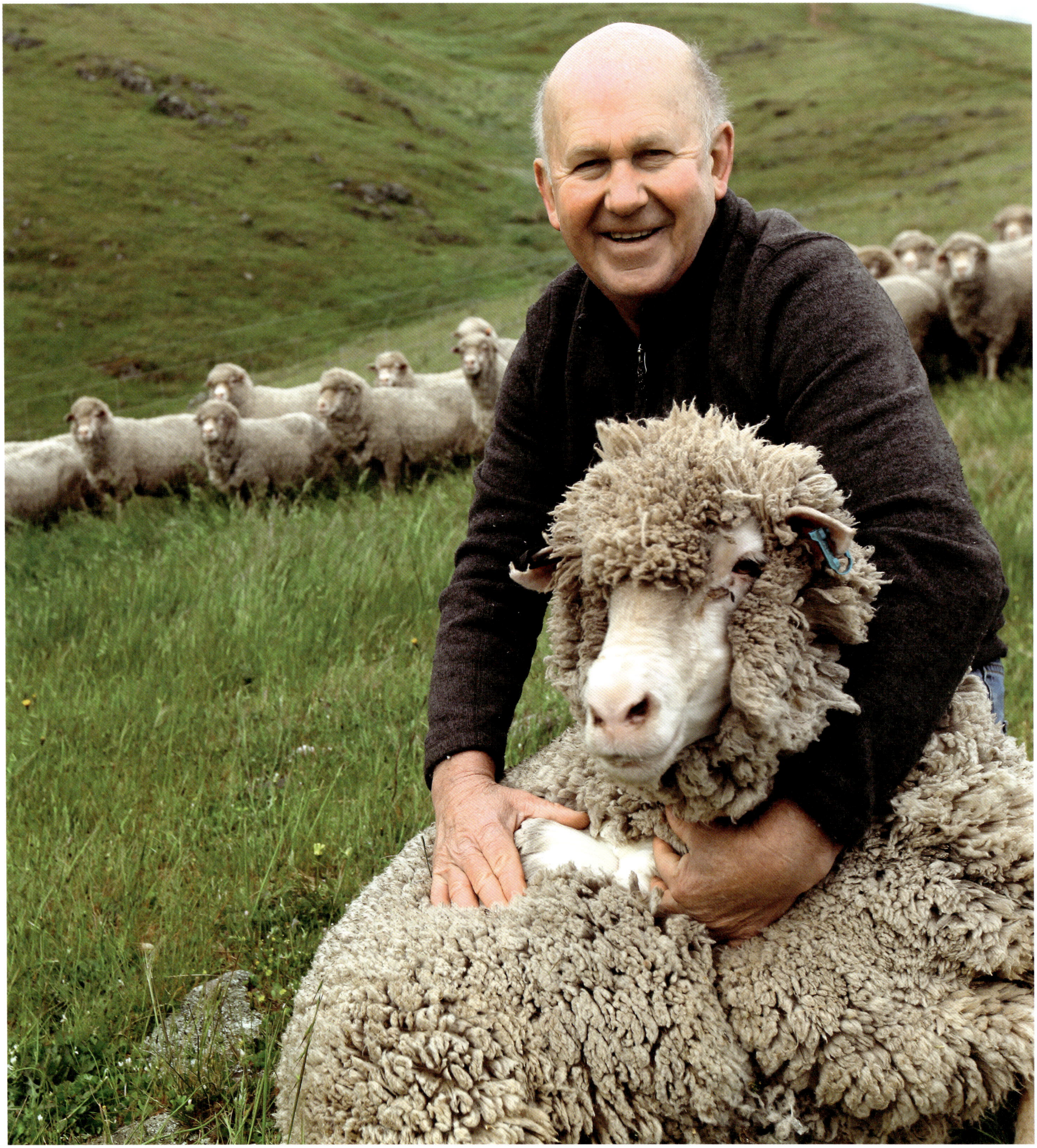

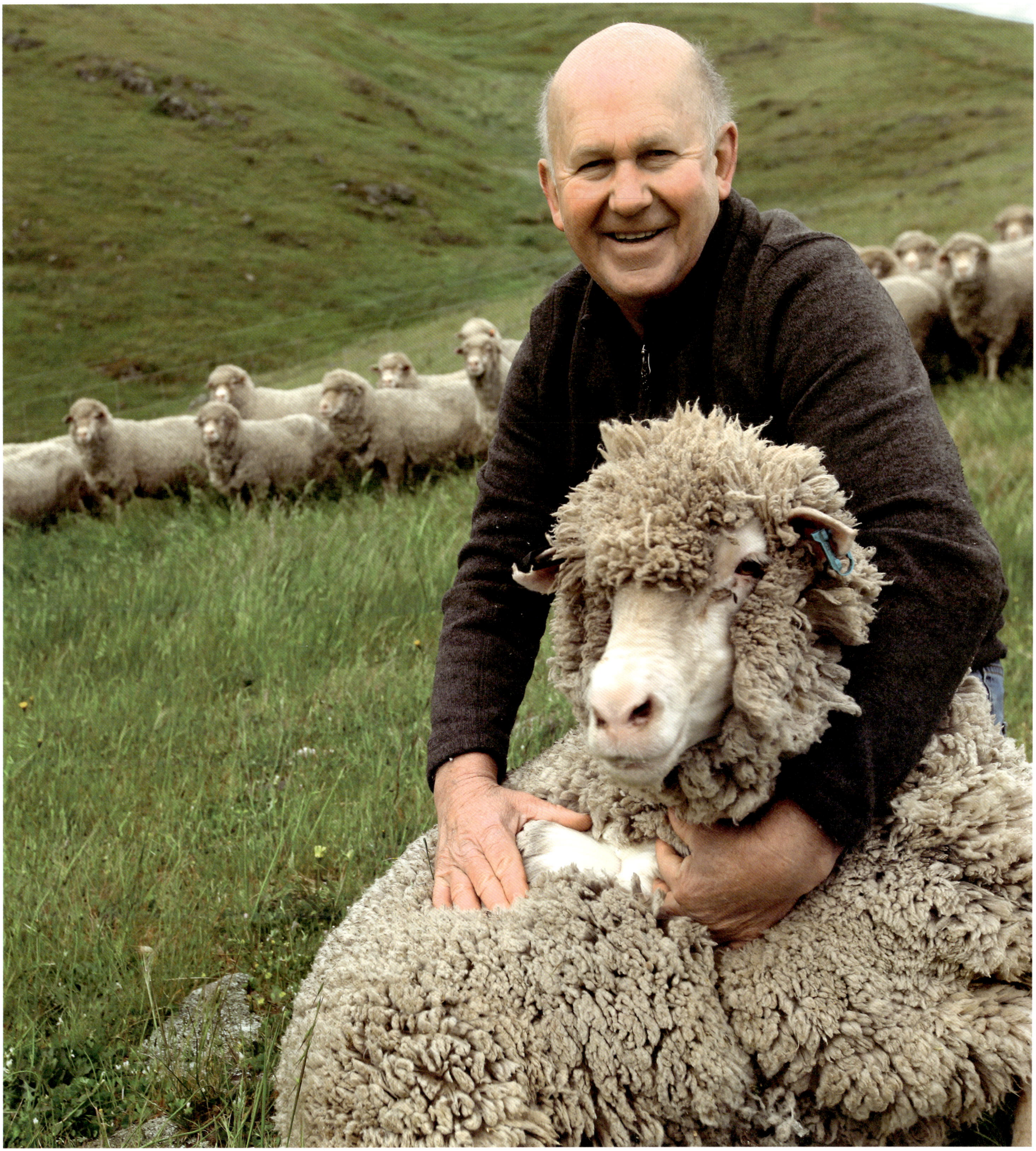

SHEARERS IN ARMIDALE, AUSTRALIA: SHEARING IS A COMPLEX OPERATION THAT CAN ONLY BE CARRIED OUT BY EXPERIENCED HANDS.

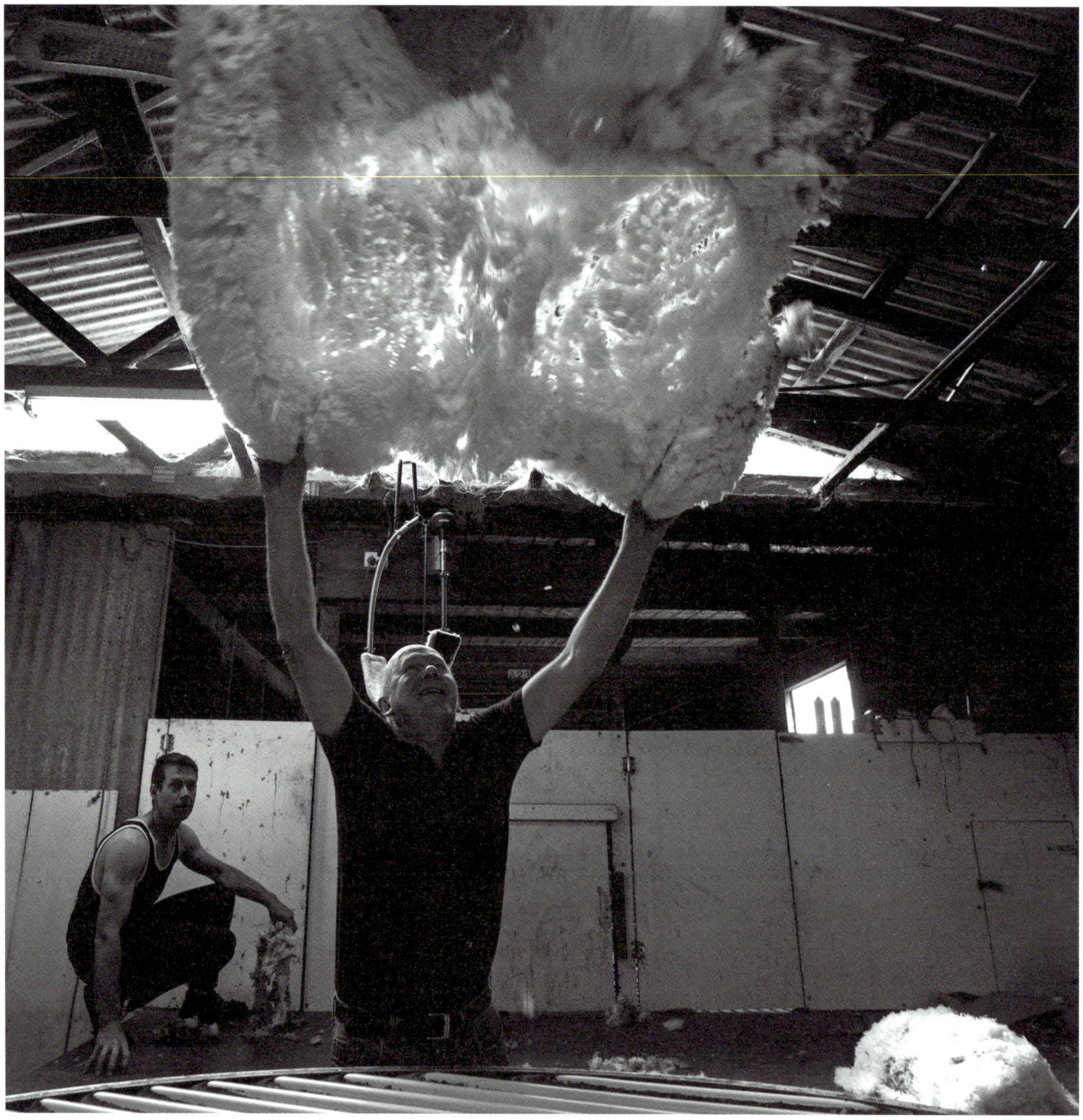

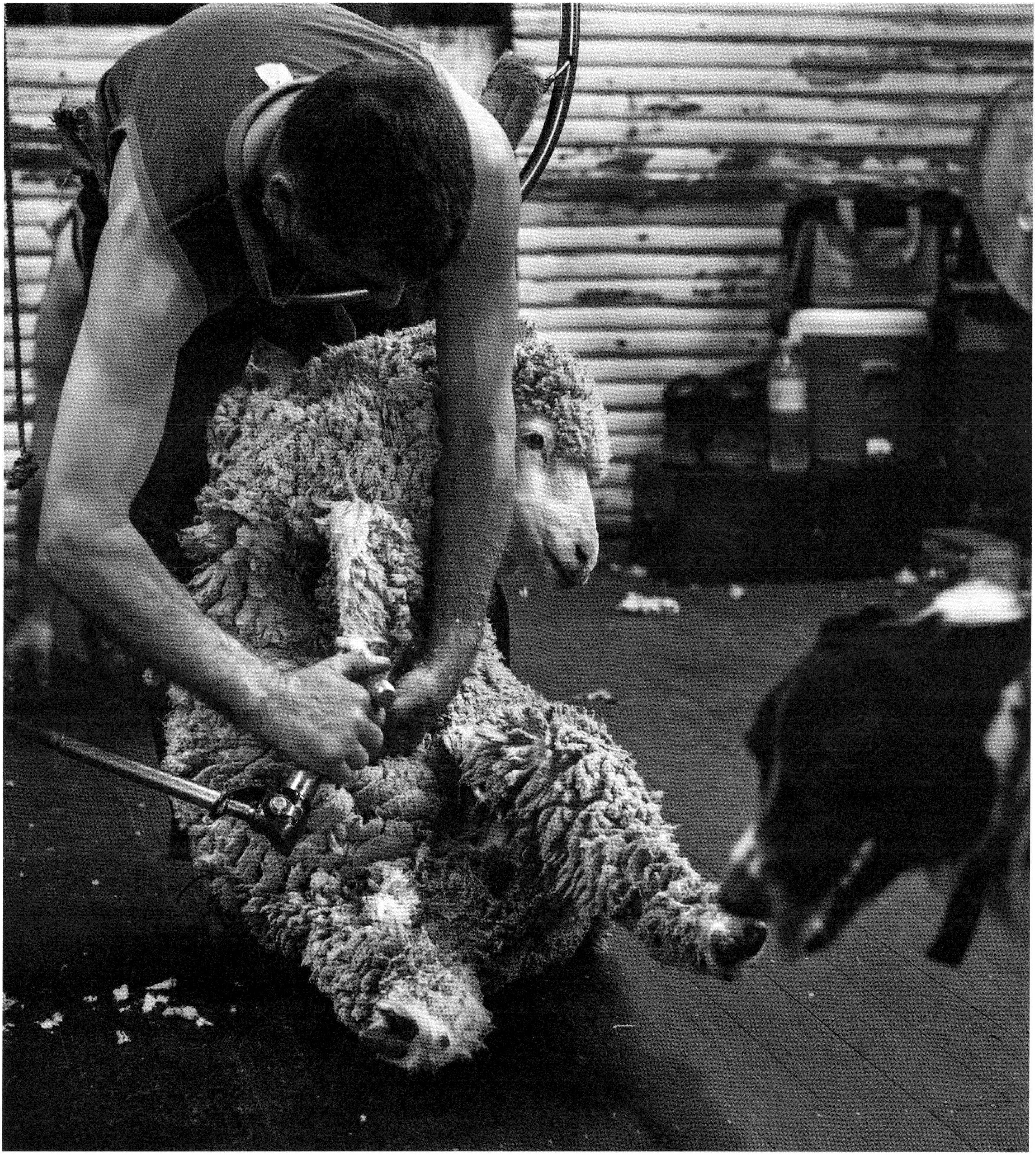

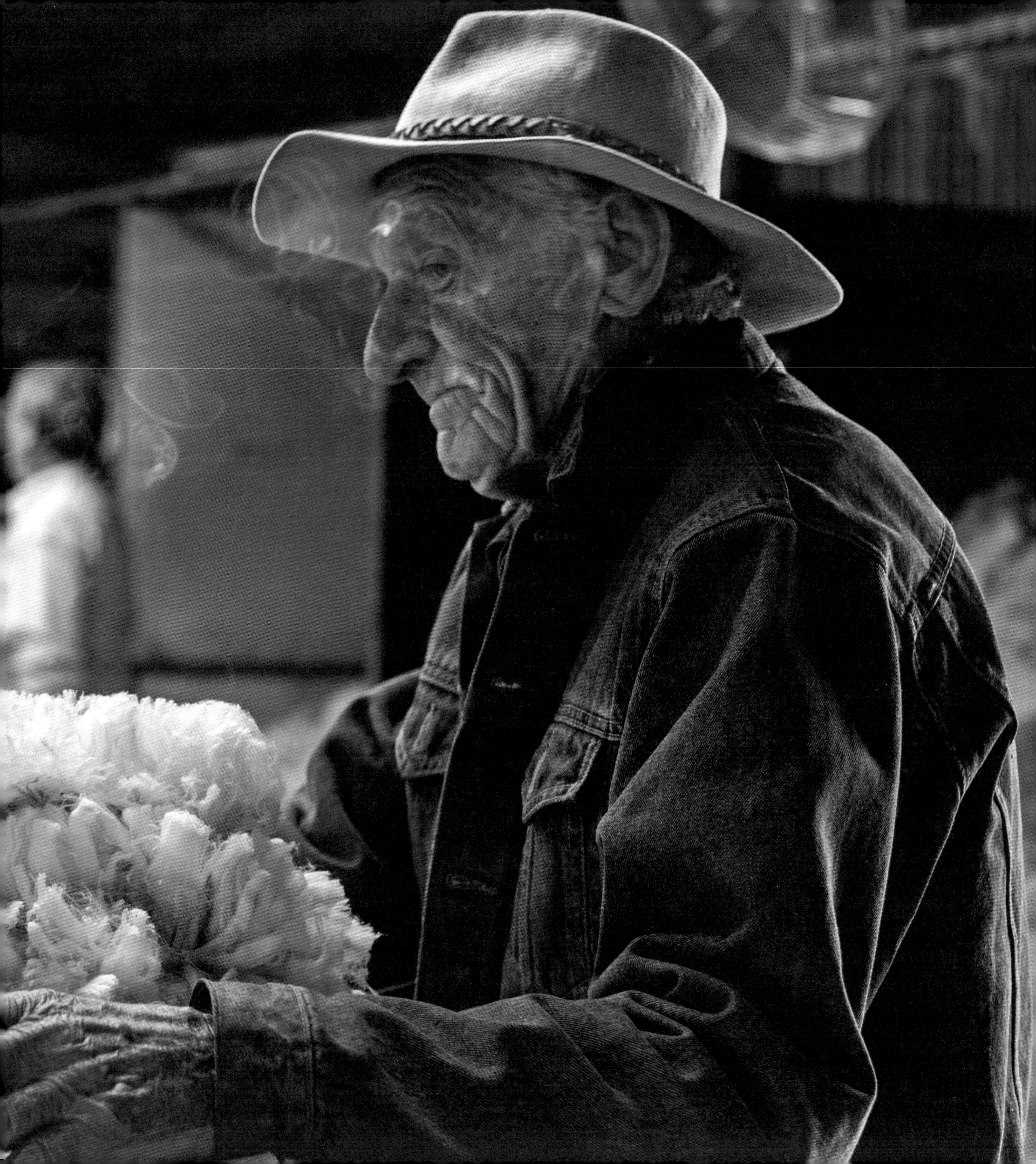

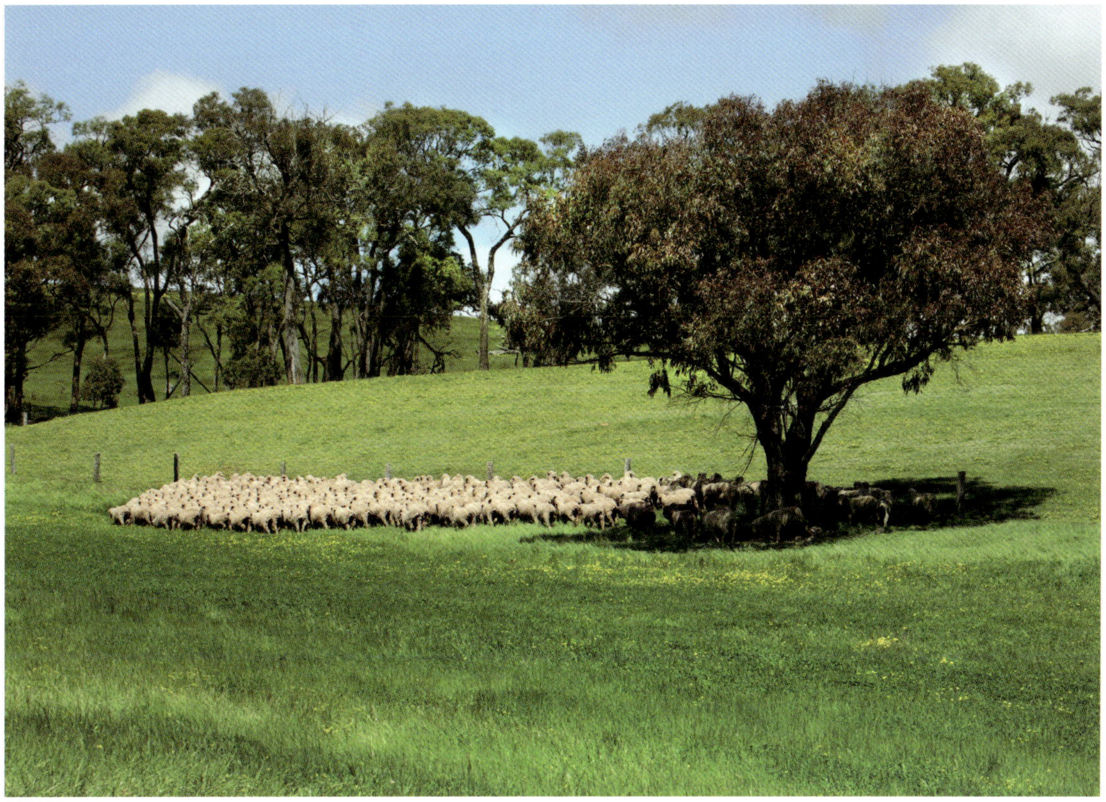

ON PYRENEES PARK FARM, AUSTRALIA, ROBERT SANDLANT PROUDLY SHOWS THE SHEEP'S MARVELLOUS FLEECE: A GENUINE GIFT OF KINGS.

Each grower has a story, each farm has its own methods, passed down through the generations. But they all share the knowledge that by taking care of, monitoring and respecting the animals and the places they live in, the wool they produce is very special.

Part that compact grey coat and underneath you will find a fleece of the purest white, silky textured and ultrasoft. Once shorn, the fibre is gossamer light, downy and smooth with a lustrous sheen and a crimped texture that lends it its natural stretch.

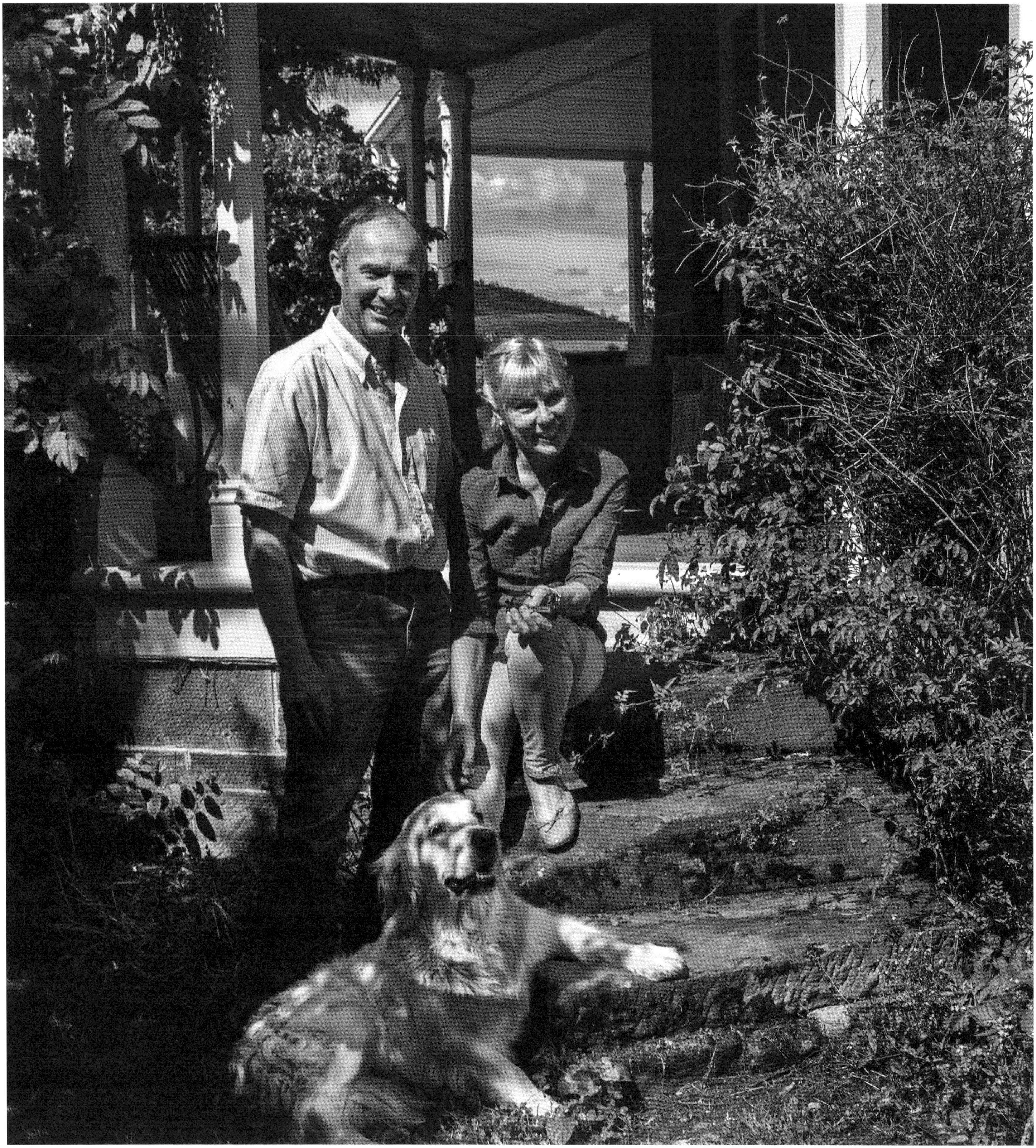

THE FARMERS MONITOR THEIR ANIMALS WITH A TRAINED, WATCHFUL, UNERRING EYE. THEIR EXPERIENCE AND DEDICATION MEAN THAT ONE LOOK AT A SHEEP'S PHYSICAL CONDITIONS ENABLES THEM TO IMMEDIATELY ASSESS ITS STATE OF HEALTH AND NEEDS.

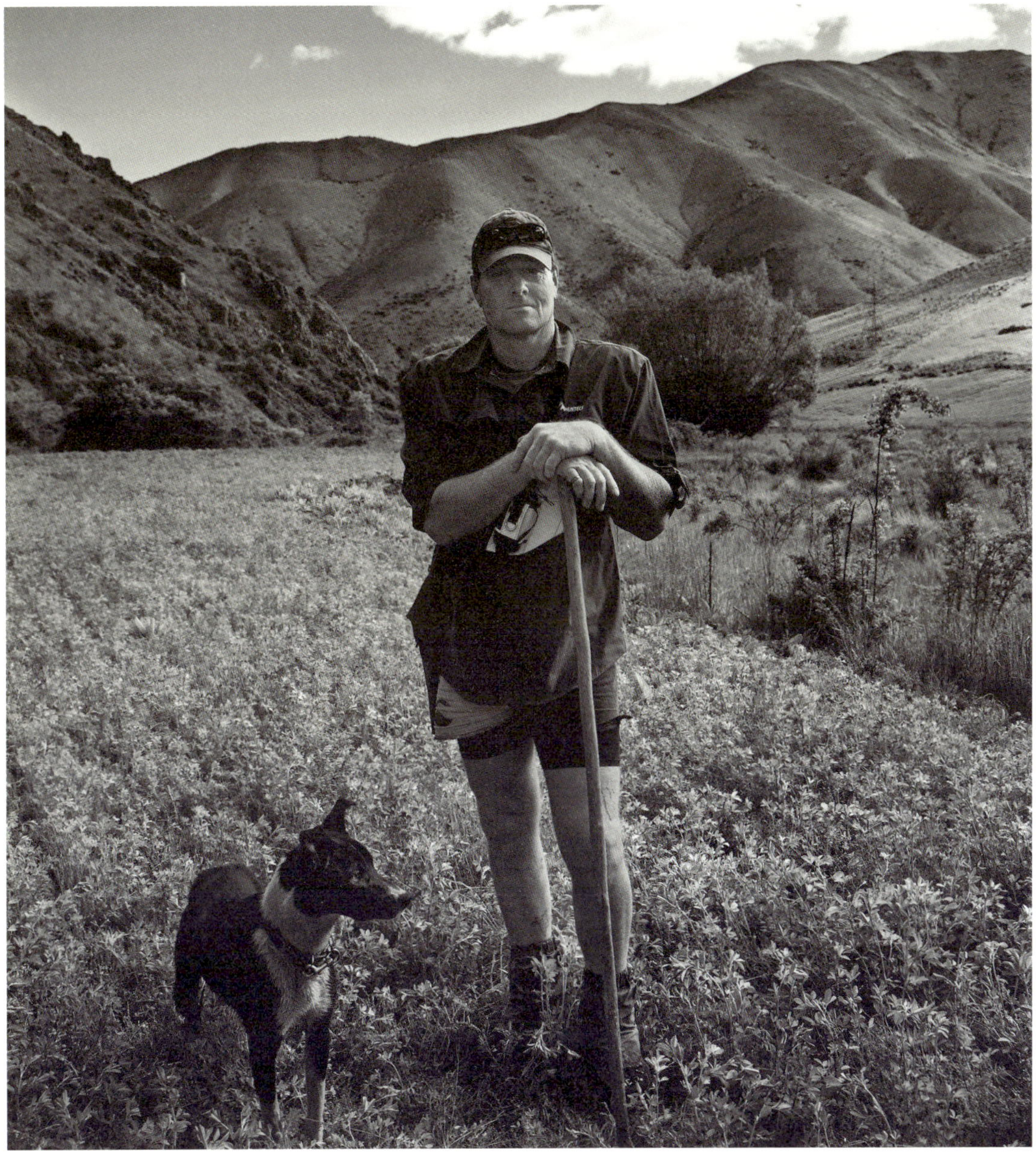

Gerald and Sue Ellis on their farm in Glenora, Tasmania and an experienced woolgrower in Black Forest, New Zealand.

following pages
The area around Lake Bolac, Dunkeld, Australia.

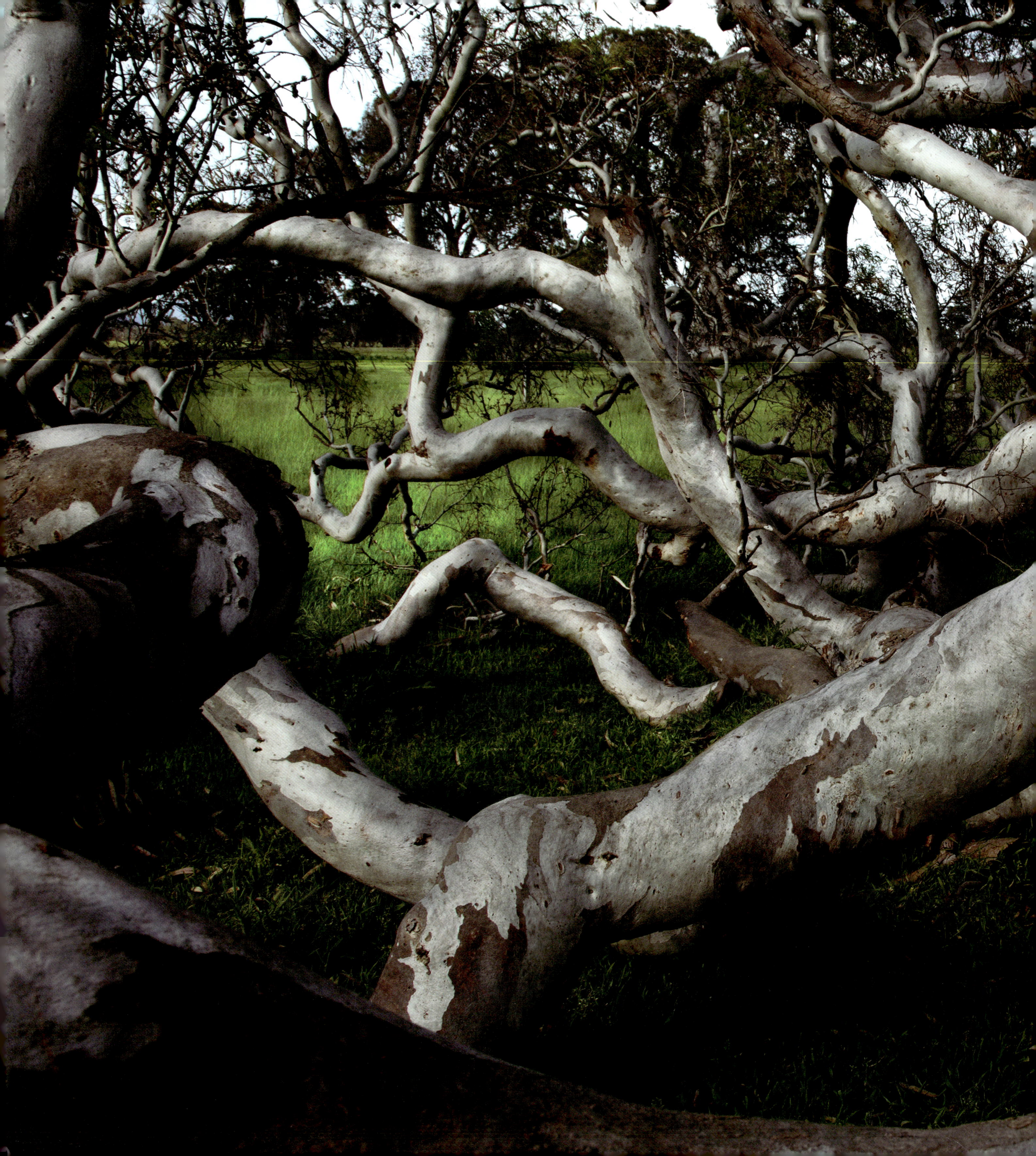

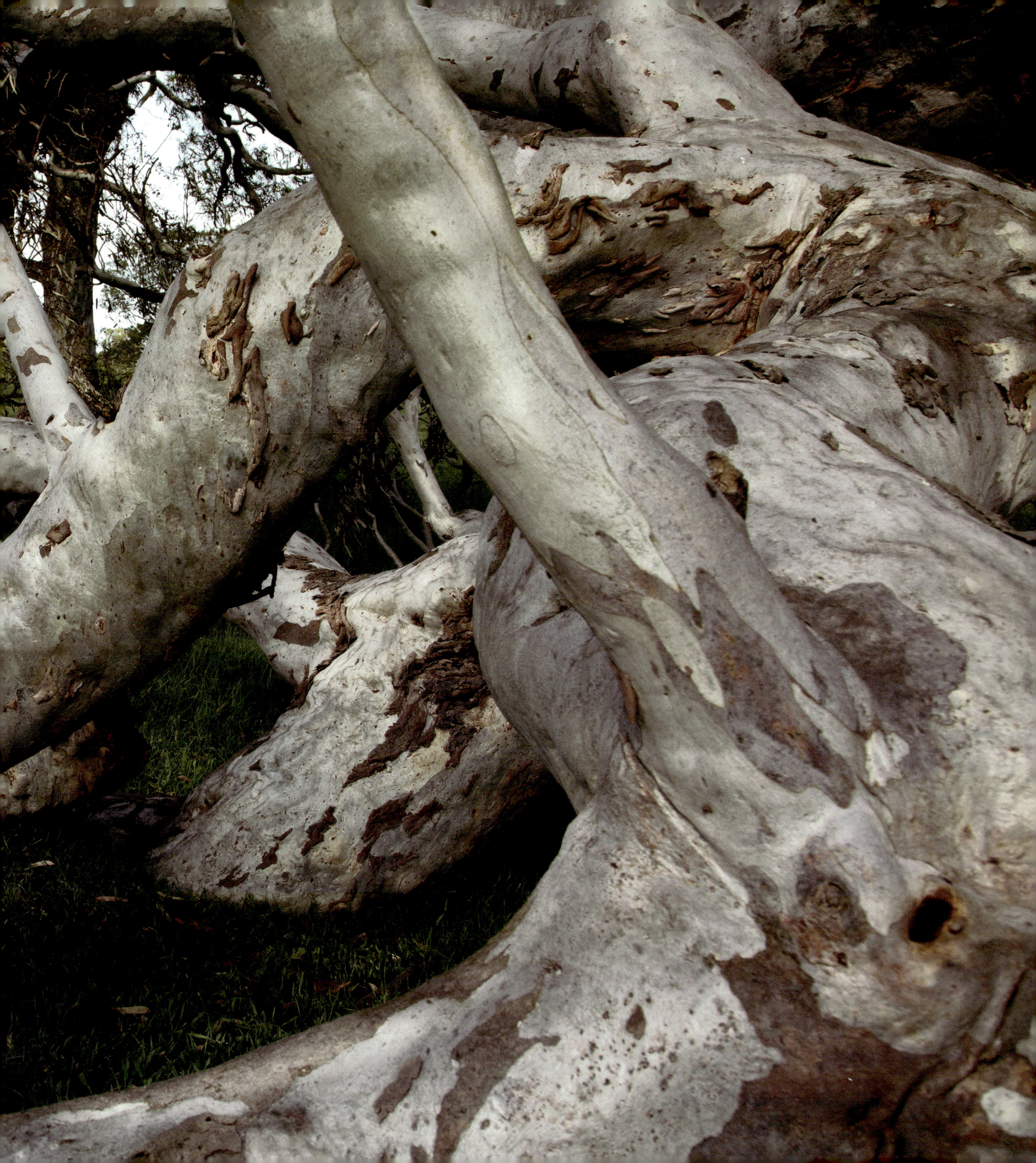

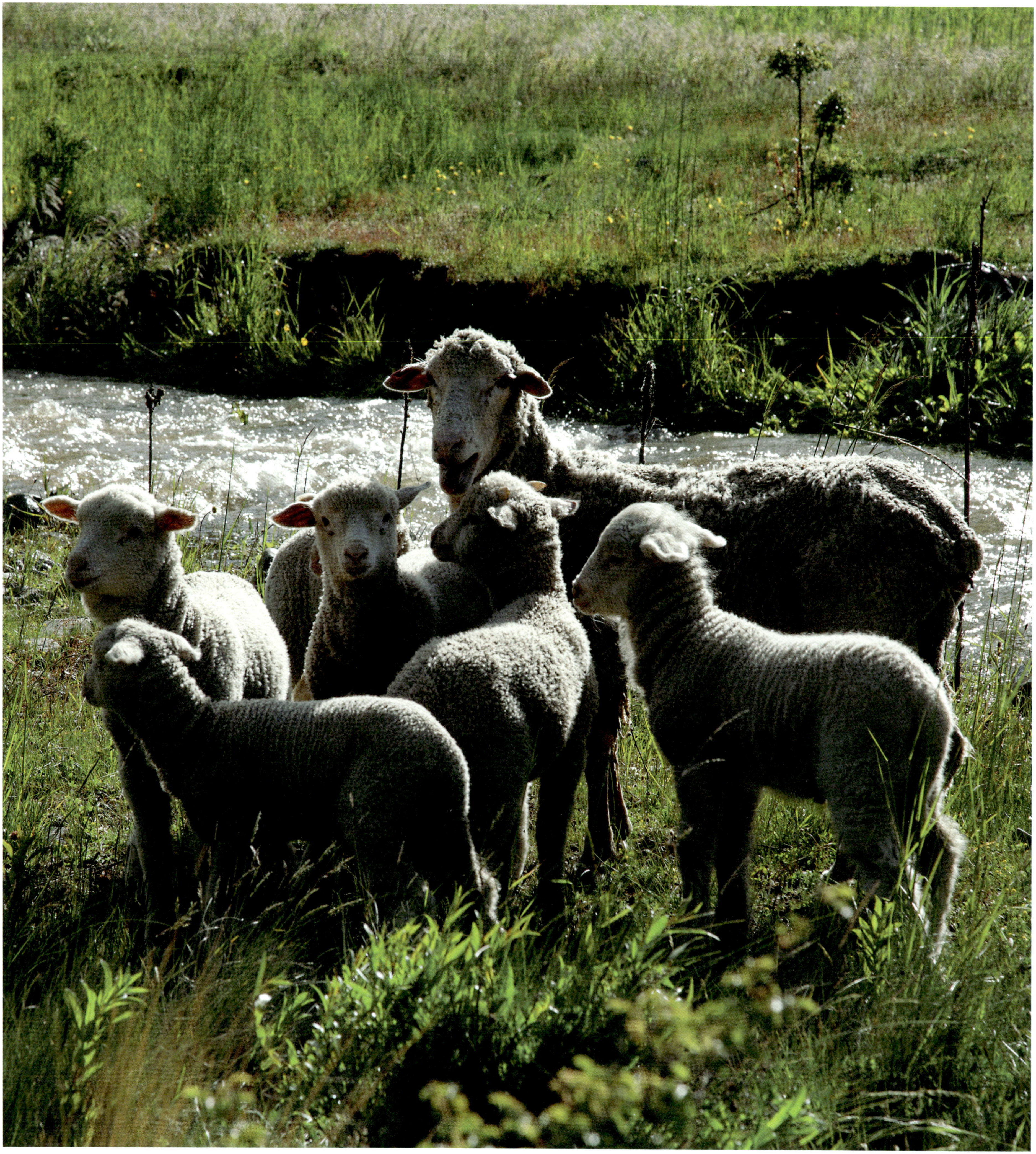

FARMING IS A WAY OF LIFE

AND A PASSION.

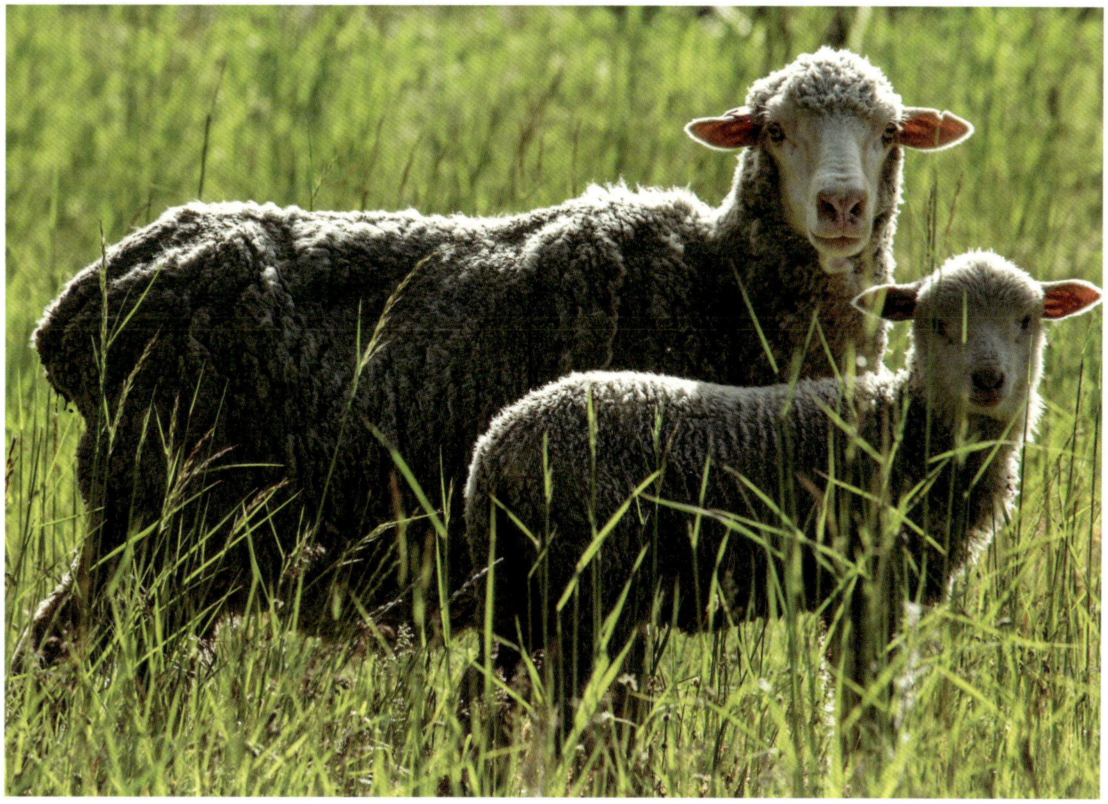

The lambs gambol on the hills, accompanied by a herder to ensure they get a healthy and varied diet. The adult sheep are constantly monitored. Sudden changes in the weather are the main worry for experienced growers: in these areas conditions can turn very rapidly, with the onset of sudden, violent rainstorms or strong winds and biting cold. Bad weather can negatively impact both on the animals' health and the quality of the fleece, but the growers are always searching new ways to protect their flocks.

Loro Piana realized it was looking at something more than just a few kilos of excellent fibre: this was a heritage, a culture, an ancient wisdom to support and cultivate.

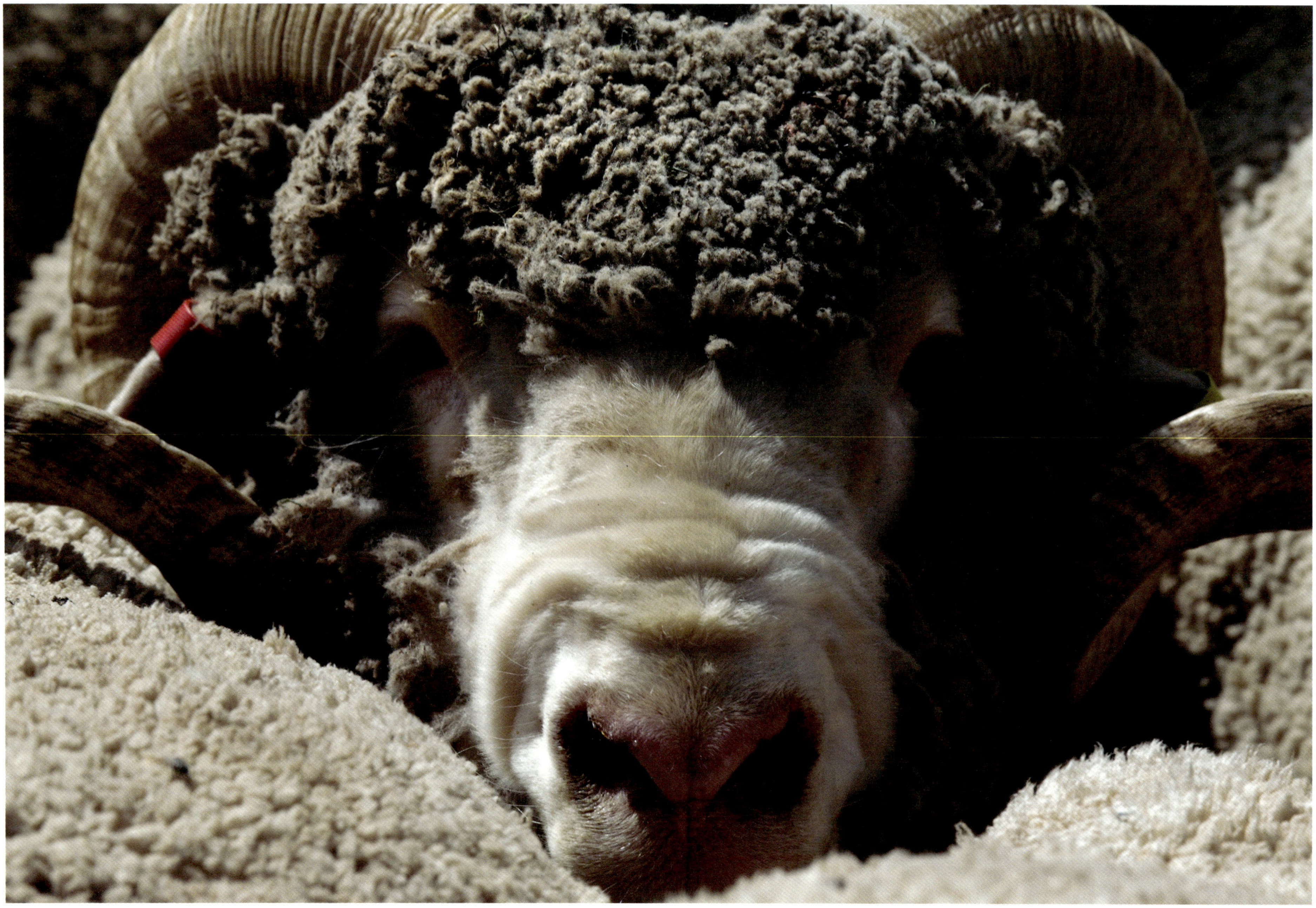

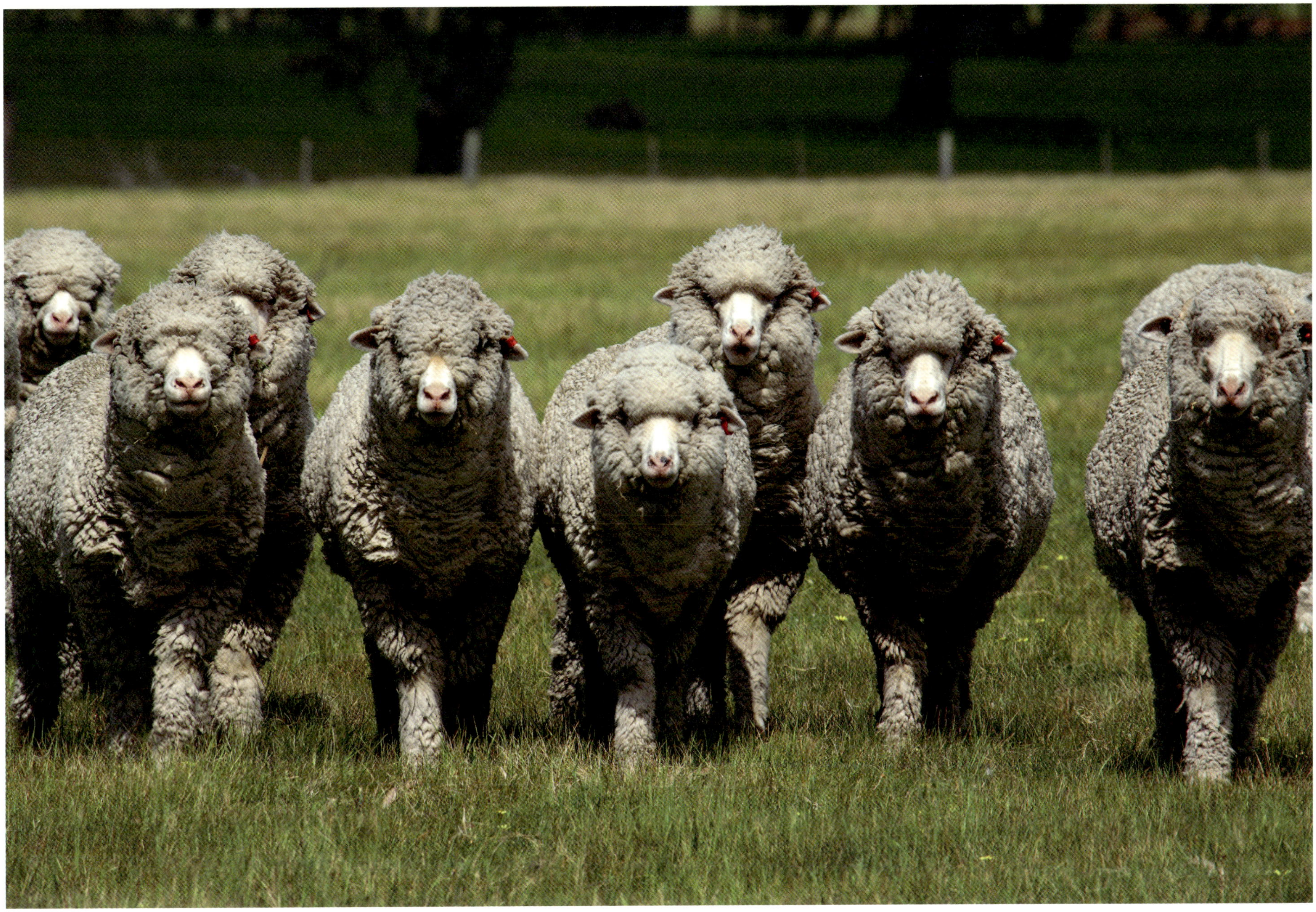

Franco Loro Piana's warning came to mind once more: the future of excellence, of this extrafine wool was in jeopardy. The growers had the same concern. They did not know how much longer they would be able to care for their animals with the same dedication, how much longer they would be able to preserve these places. One farmer in particular was to lead the way for Loro Piana. His name was Donald Burnett and he was a New Zealander. In his eyes, on his face, in his words was the story of these animals, the passion involved in looking after them. Burnett loved his land and saw his lifestyle as a treasure to cherish: a gift of destiny, being chosen to live in these beautiful places, at one with nature and nurturing her precious resources.

Donald Burnett was the spiritual heir of generations of enlightened farmers, merchants, explorers and sovereigns who, since ancient times, with bold vision and a passion for quality, have ensured the survival and spread of this breed of sheep throughout the world.

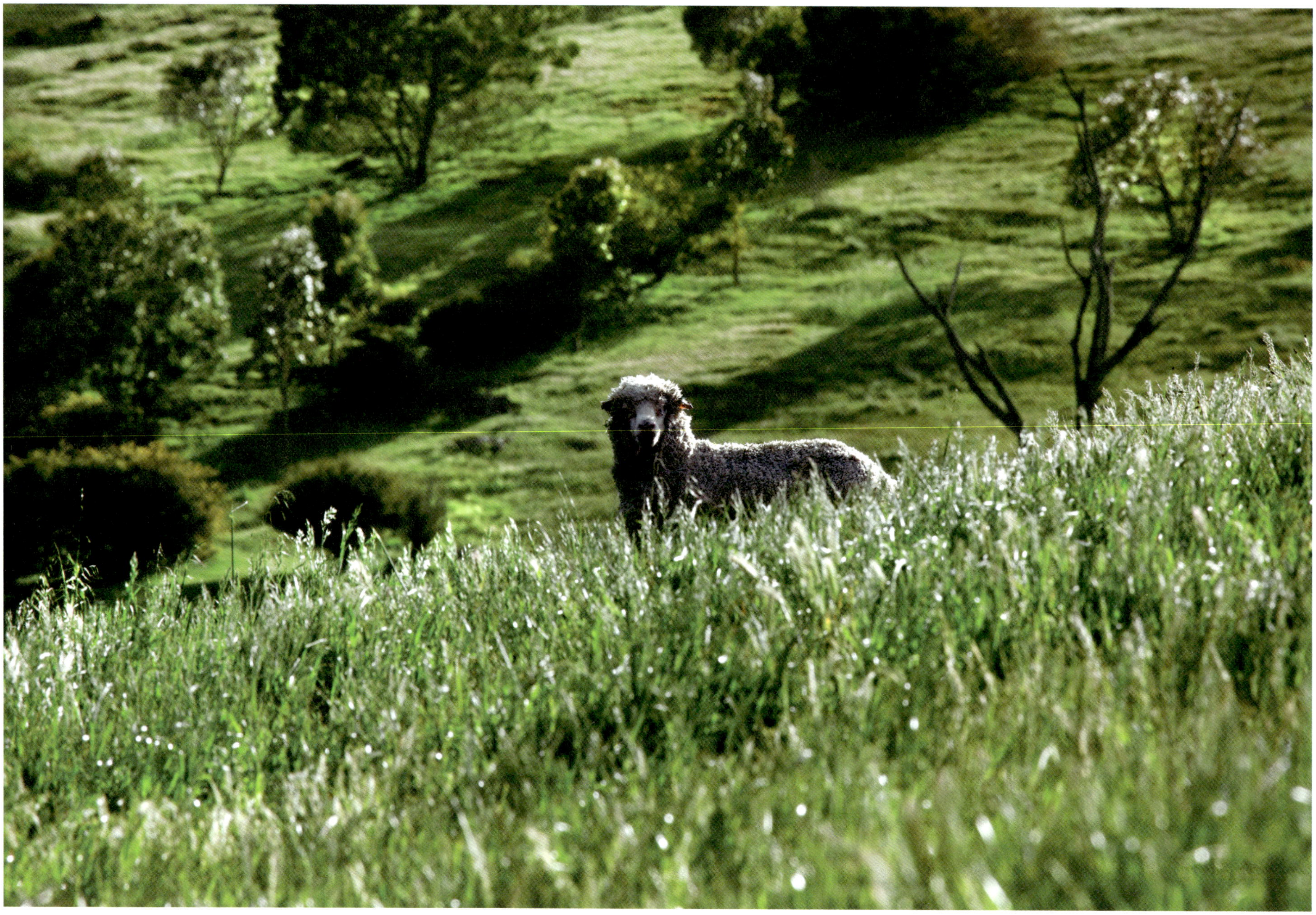

following pages
James Hume on his farm in Gretna, Tasmania.

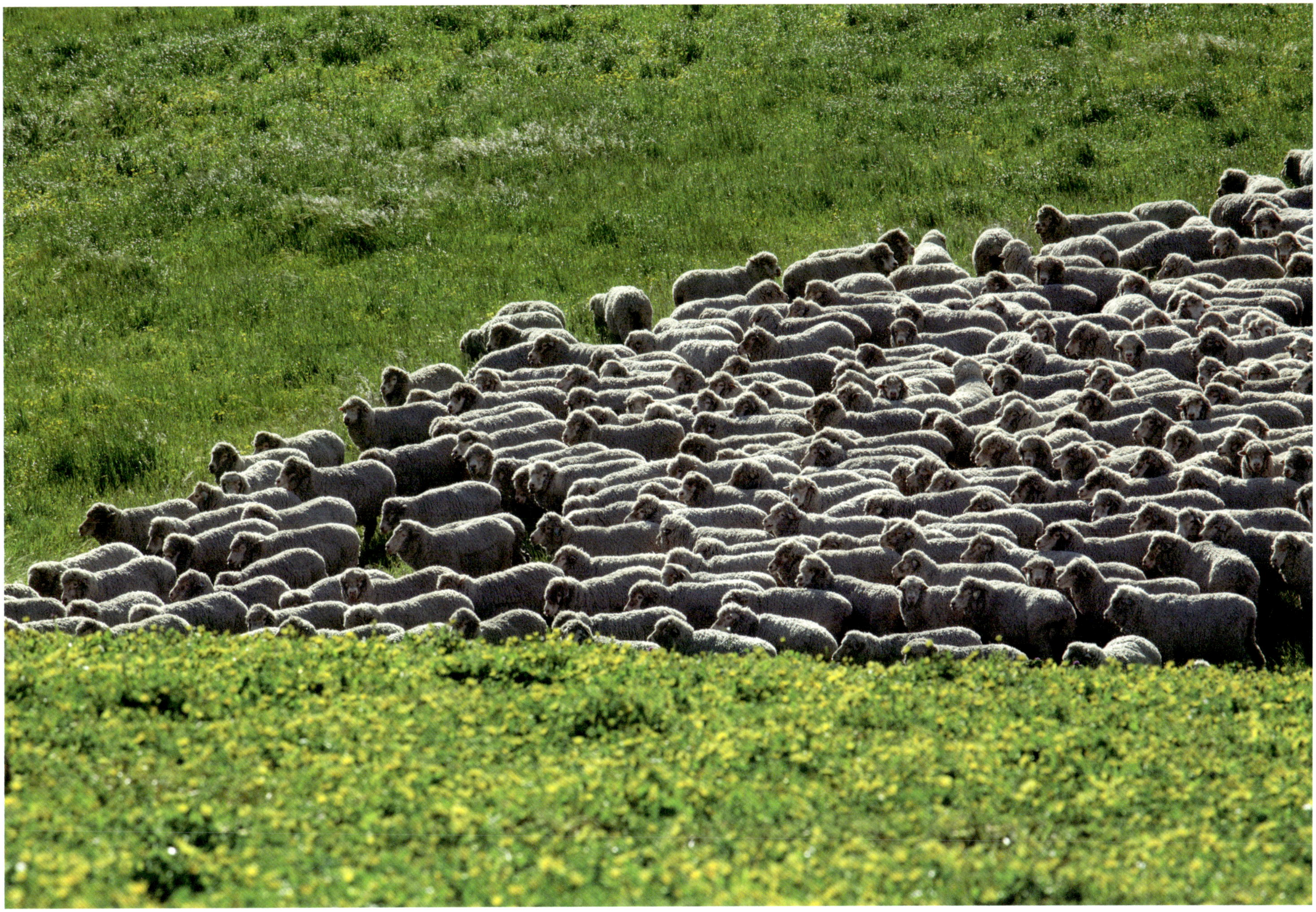

The story of the breed indeed goes a long way back: its fleece was traded by the Phoenicians, who took it to the Middle East and throughout the Mediterranean. The Islamic tribes in North Africa successfully bred merino sheep and brought them to Spain, where the climate and conditions proved particularly favourable. Their ultrafine wool soon became one of the country's most highly renowned, sought-after products.

After the Spanish Reconquista, the new sovereigns did not merely inherit this flourishing activity, but made it into something exclusive, the preserve of the upper classes. Spain's rulers traded in the wool and also used it as a gift for diplomatic purposes: a pair of merino sheep thus became "the gift of kings" destined for the most illustrious recipients.

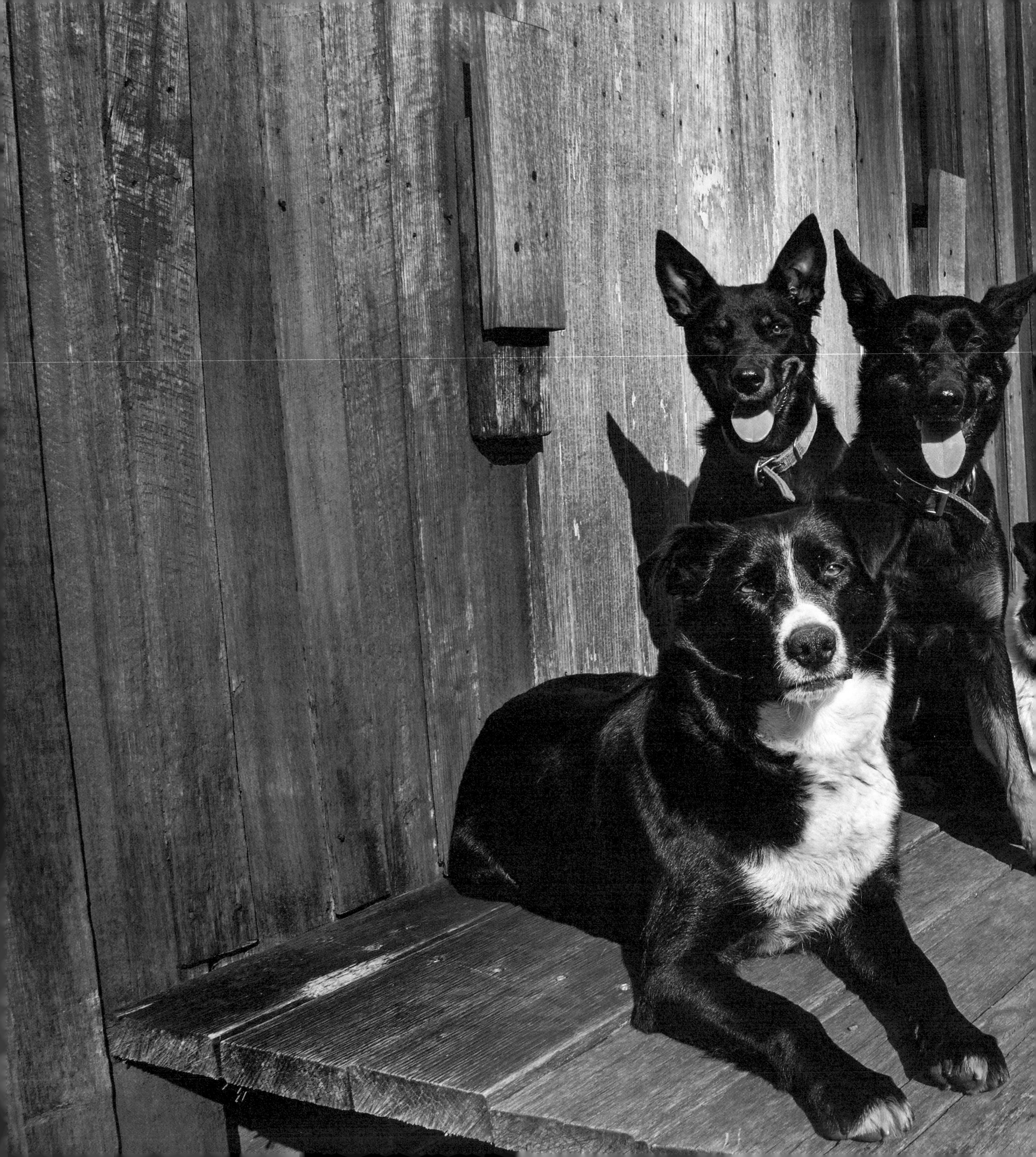

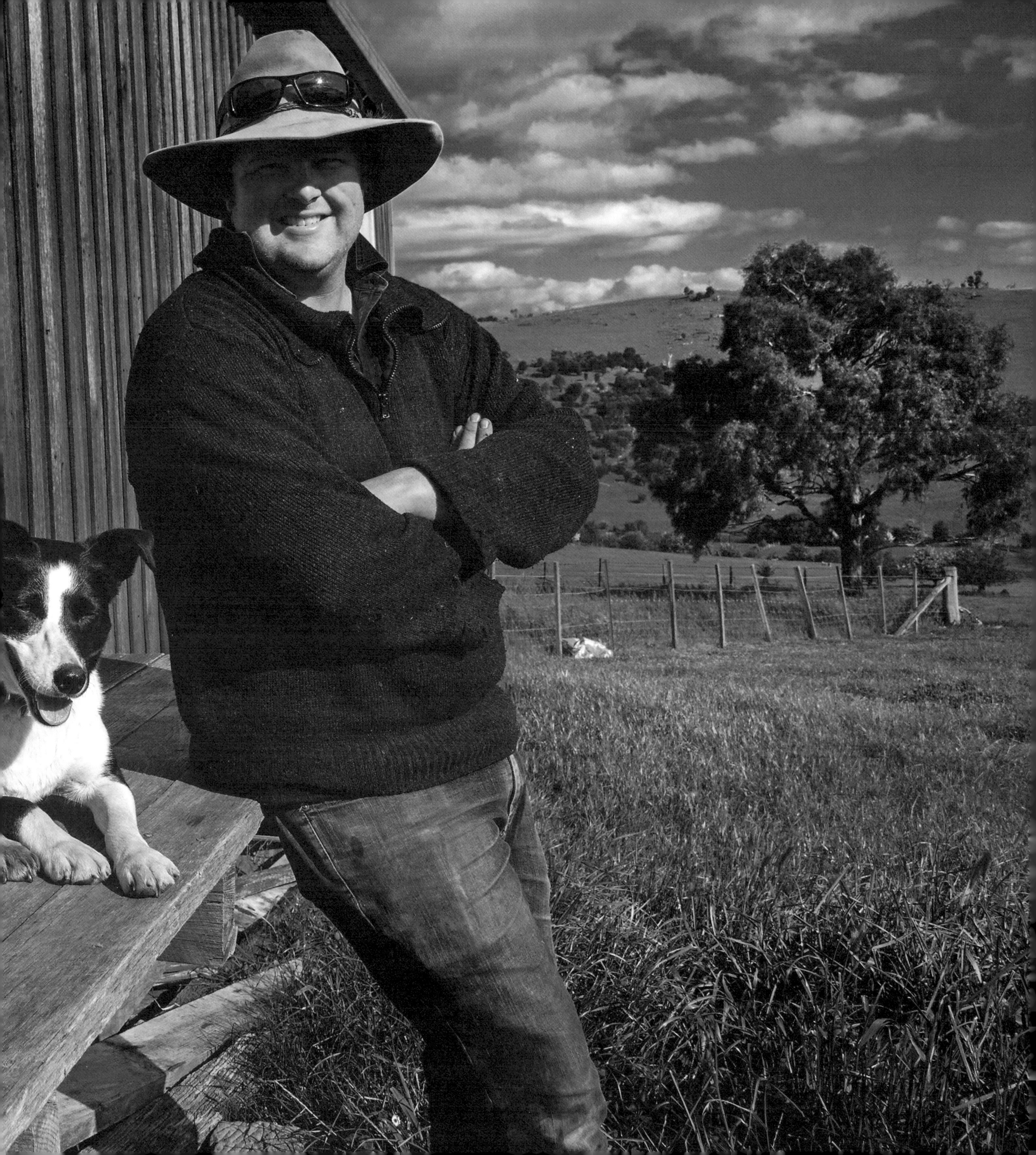

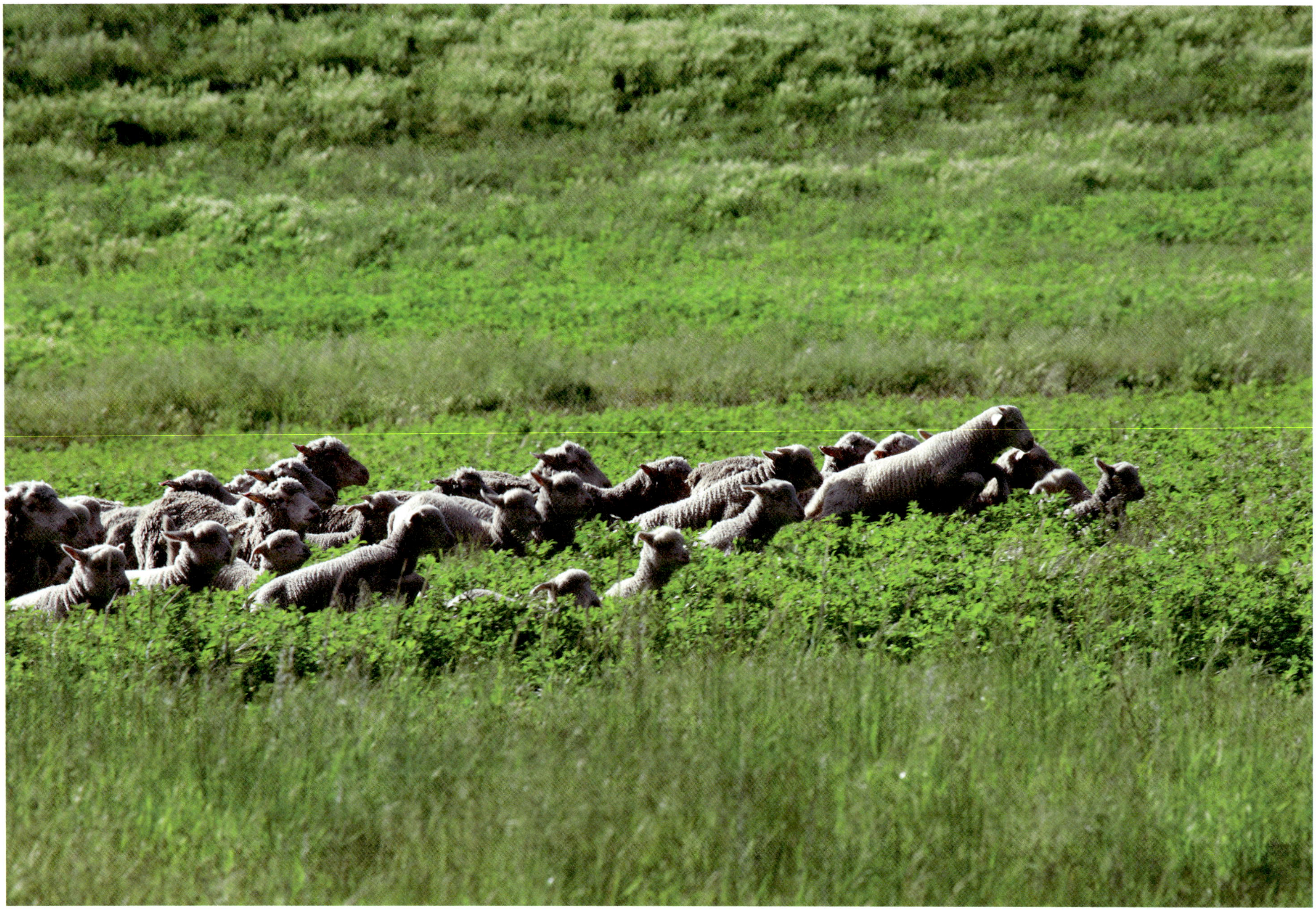

Flocks grazing in Australia:
areas of Victoria and Dunkeld.

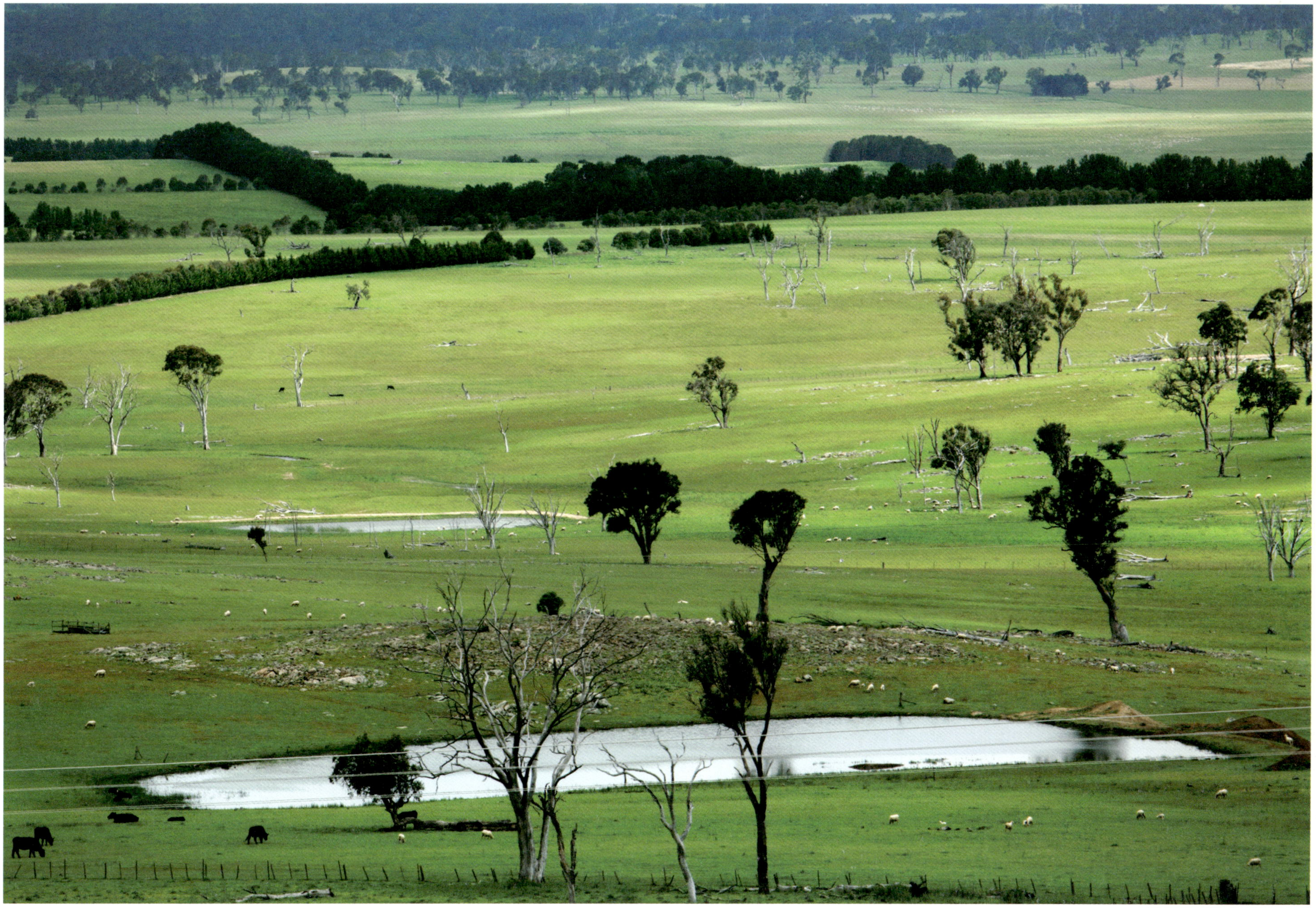

Thanks to these placid white sheep, pacts were sealed, alliances formed, and the political future of Europe was sketched out. The gift to the Elector of Saxony in 1765 in particular took this precious breed to pastures new, where it was put to good use and efforts were made to improve the breed and increase wool production. Meanwhile Europeans were engaged in voyages of exploration and conquest in distant lands.

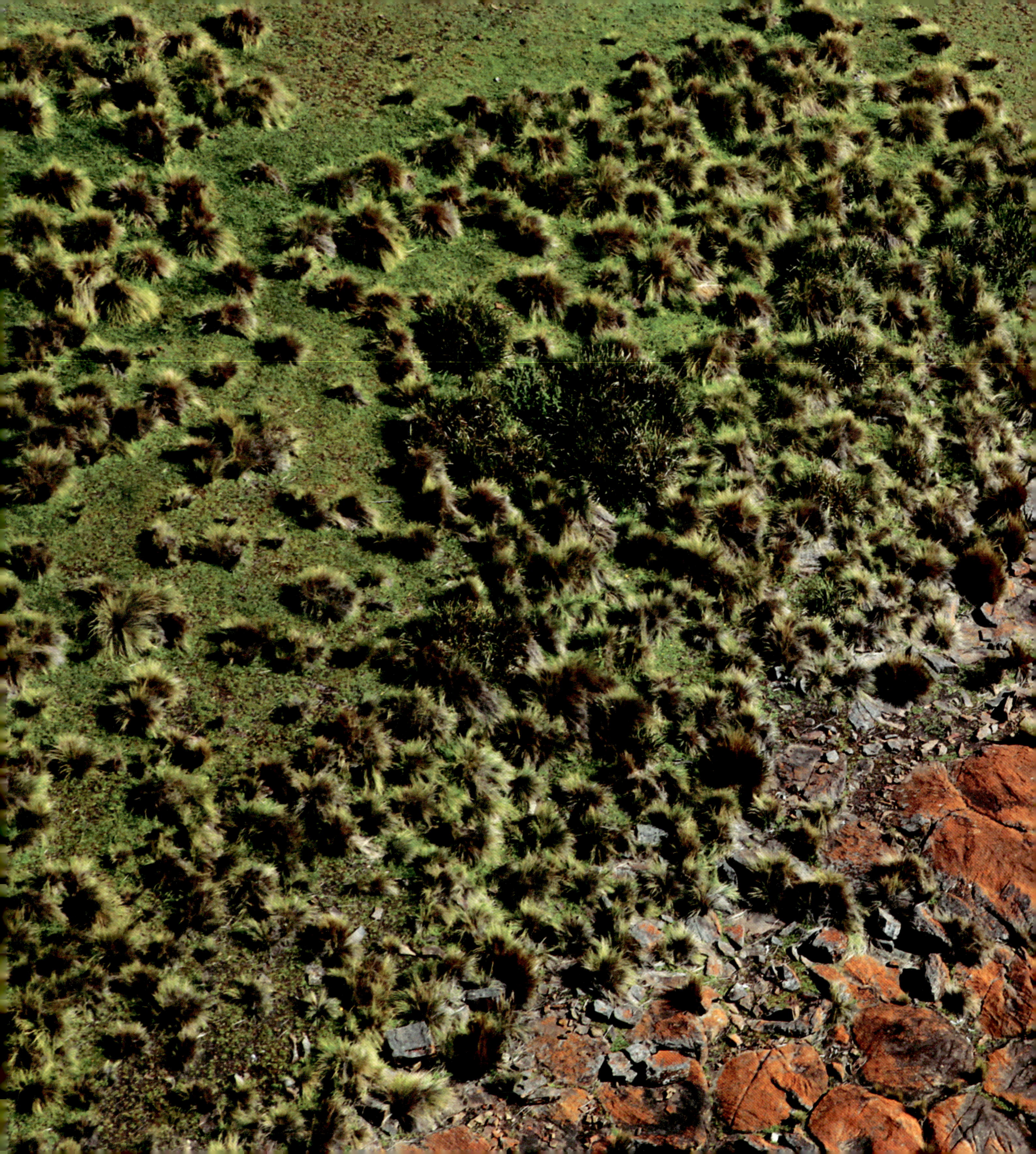

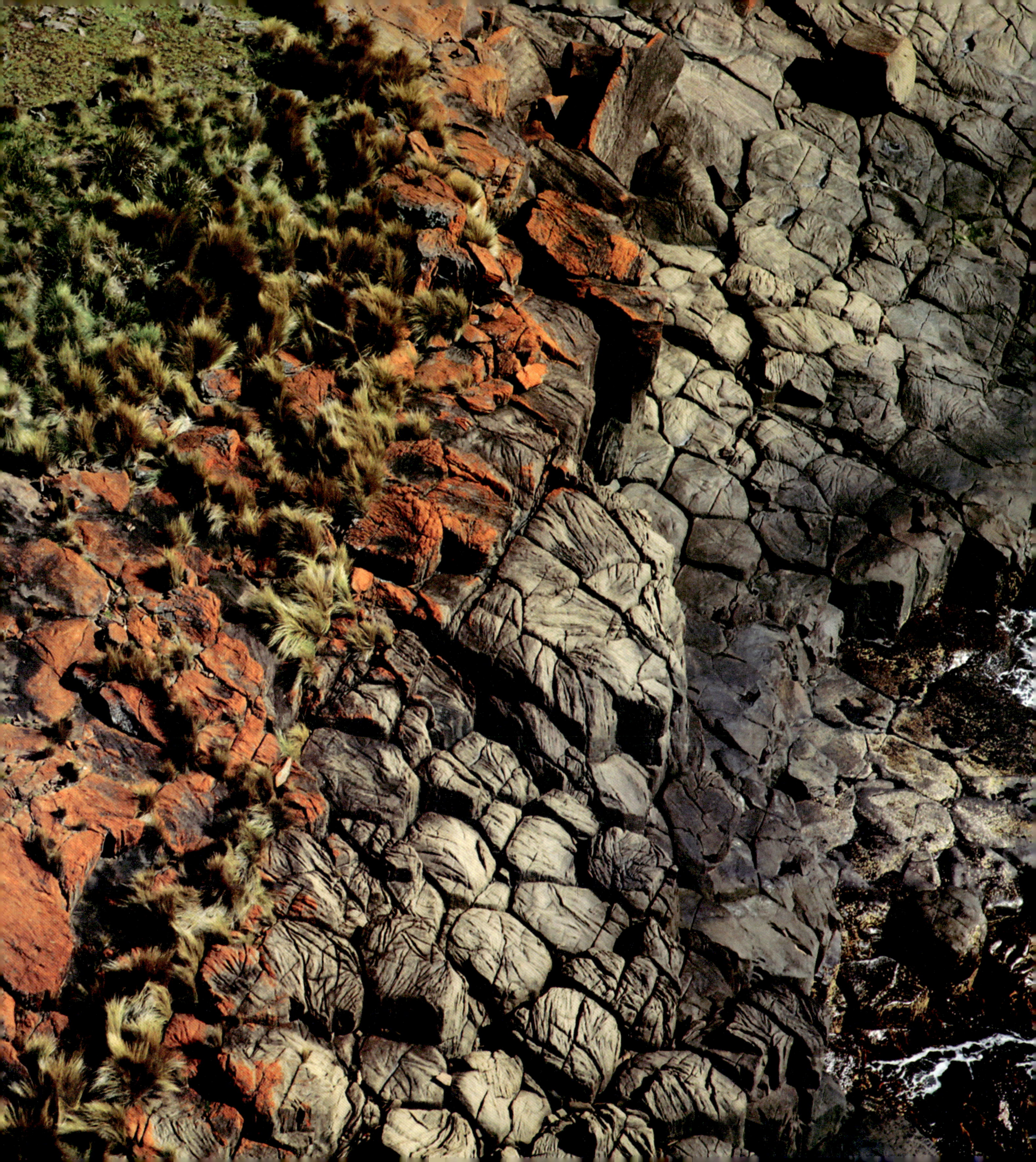

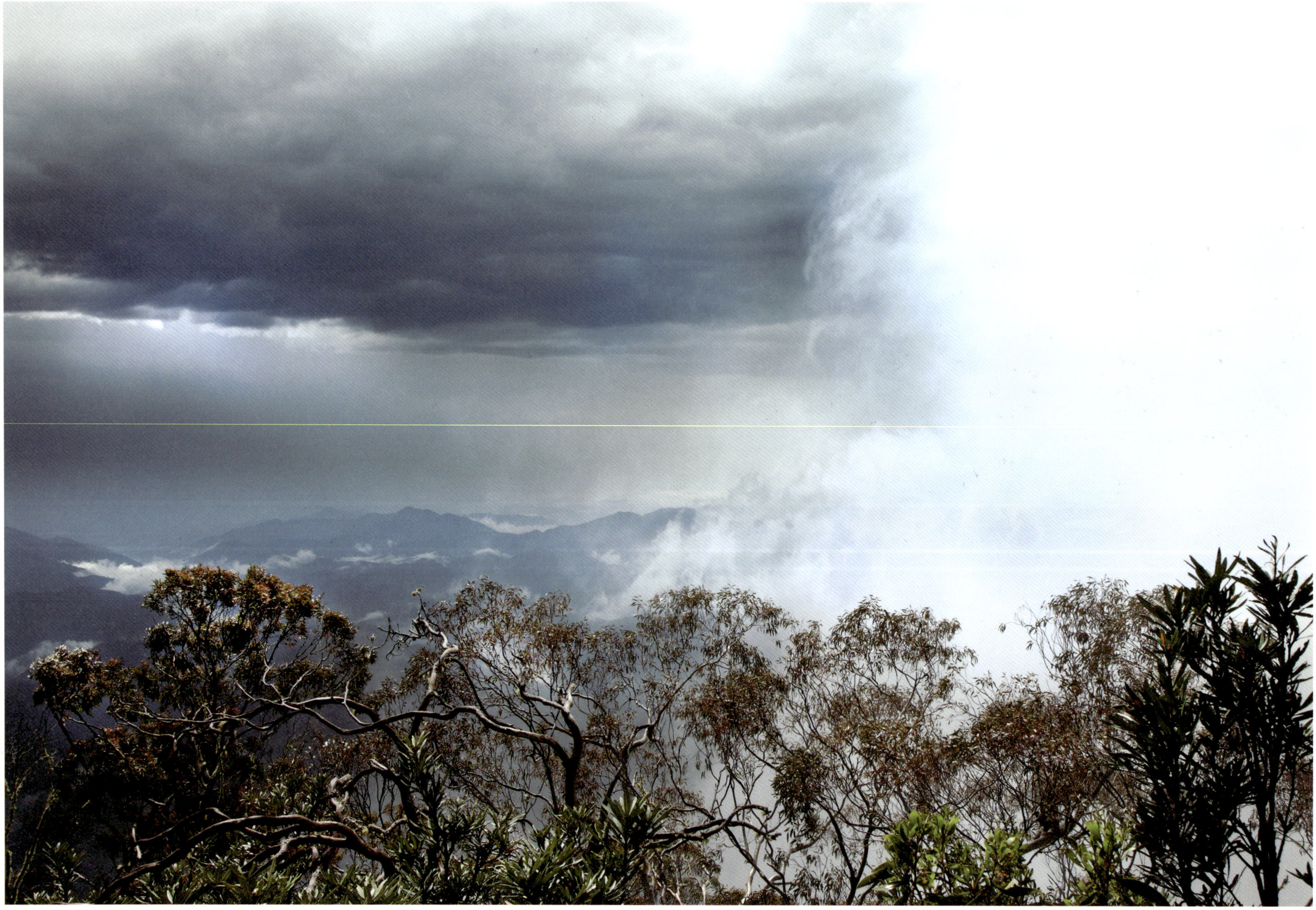

previous pages
The stunning hues and textures
of the Tasmanian coast.

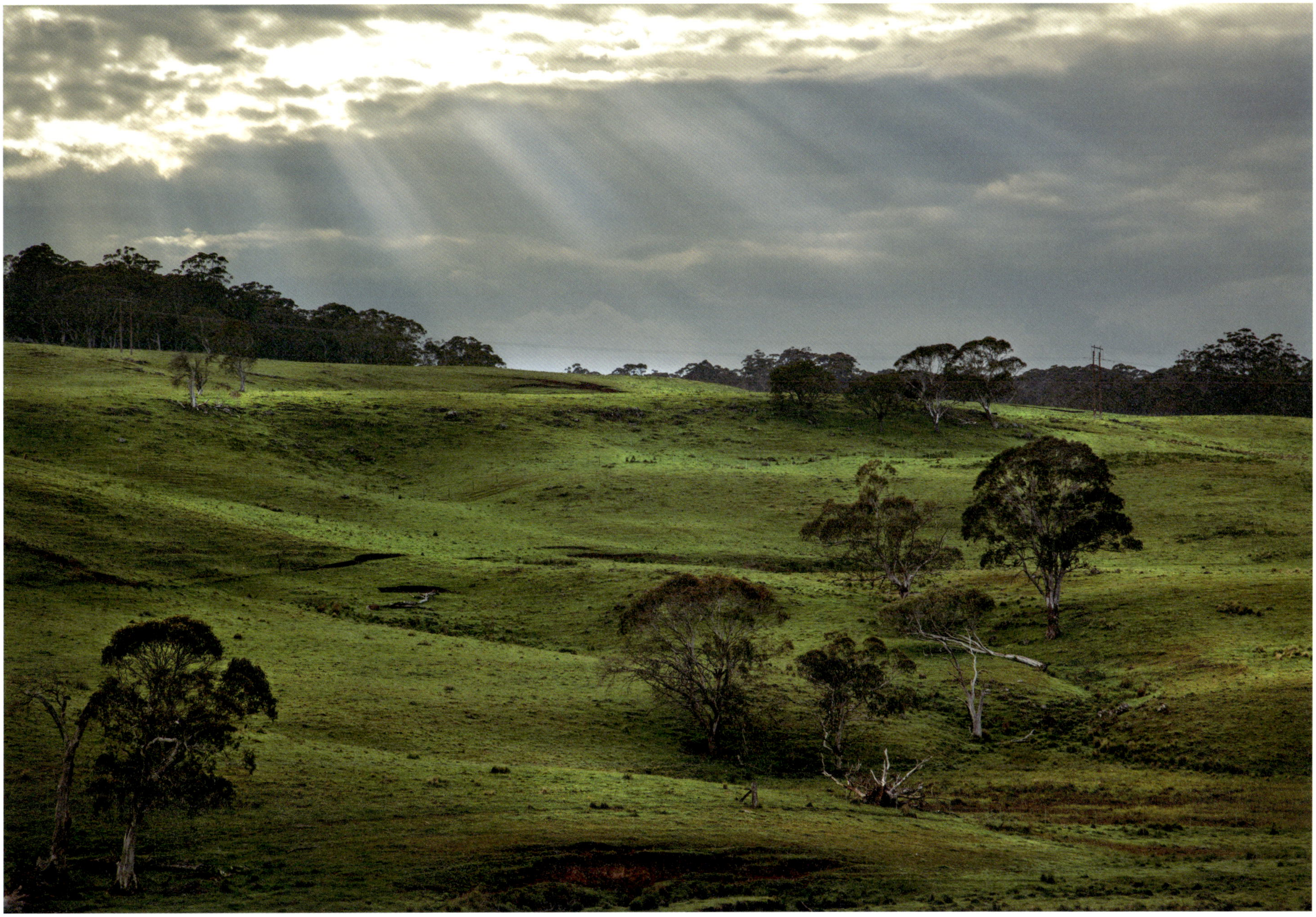

In 1773 Captain James Cook led an expedition to New Zealand. It was his second time in these seas, the second time that he set foot on these islands of lush vegetation, untouched rivers and coasts, at the edges of the earth. With him, he had a pair of merino sheep. Only a few years later, the sheep also landed in Australia, where they were pure-bred in optimal conditions: the country offered vast areas of unspoilt land and a fledgling economy. The long journey of the merino sheep seemed to have reached its ideal destination.

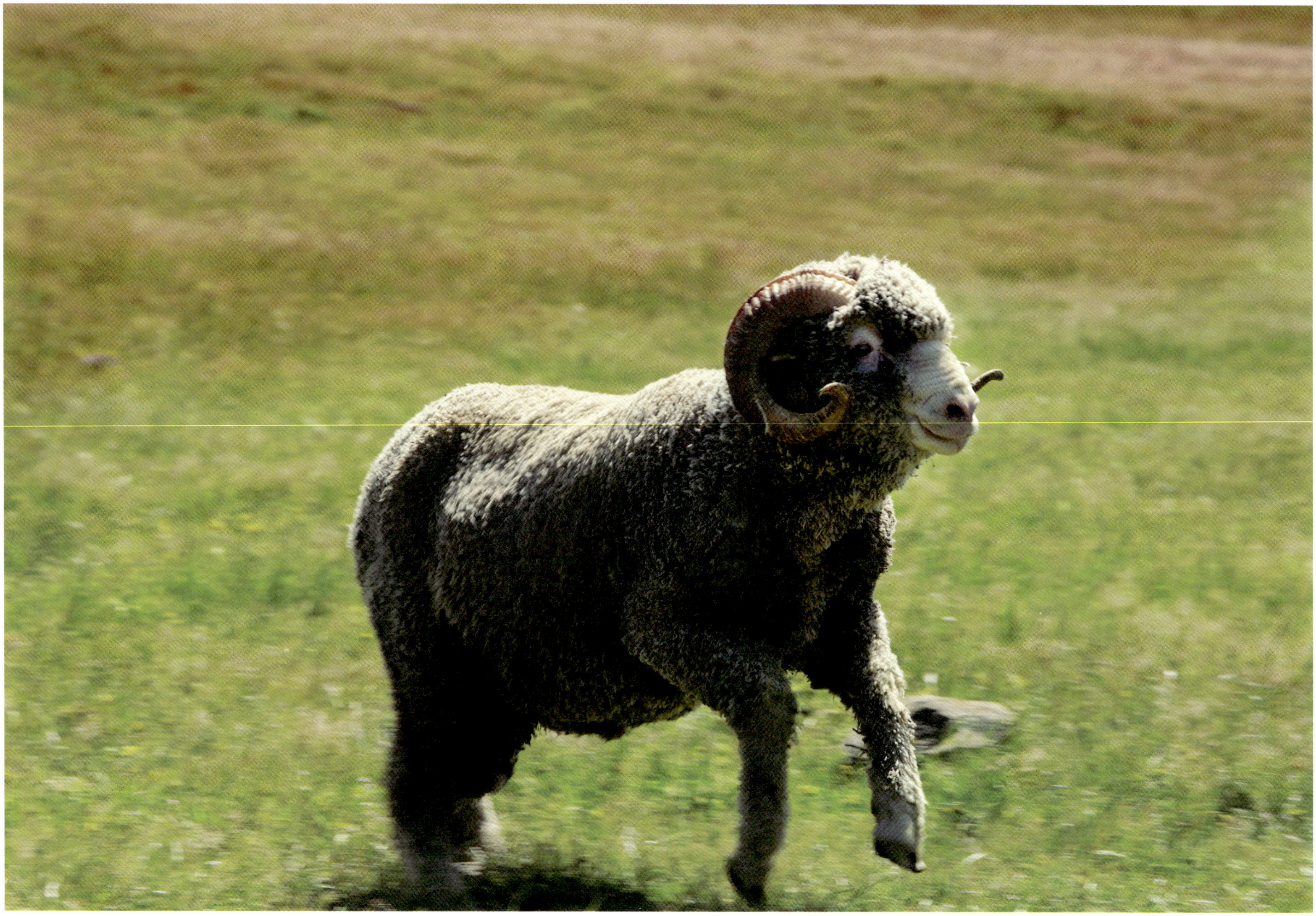

following pages
The vast pastures of Pyrenees Park,
Central Victoria, Australia.

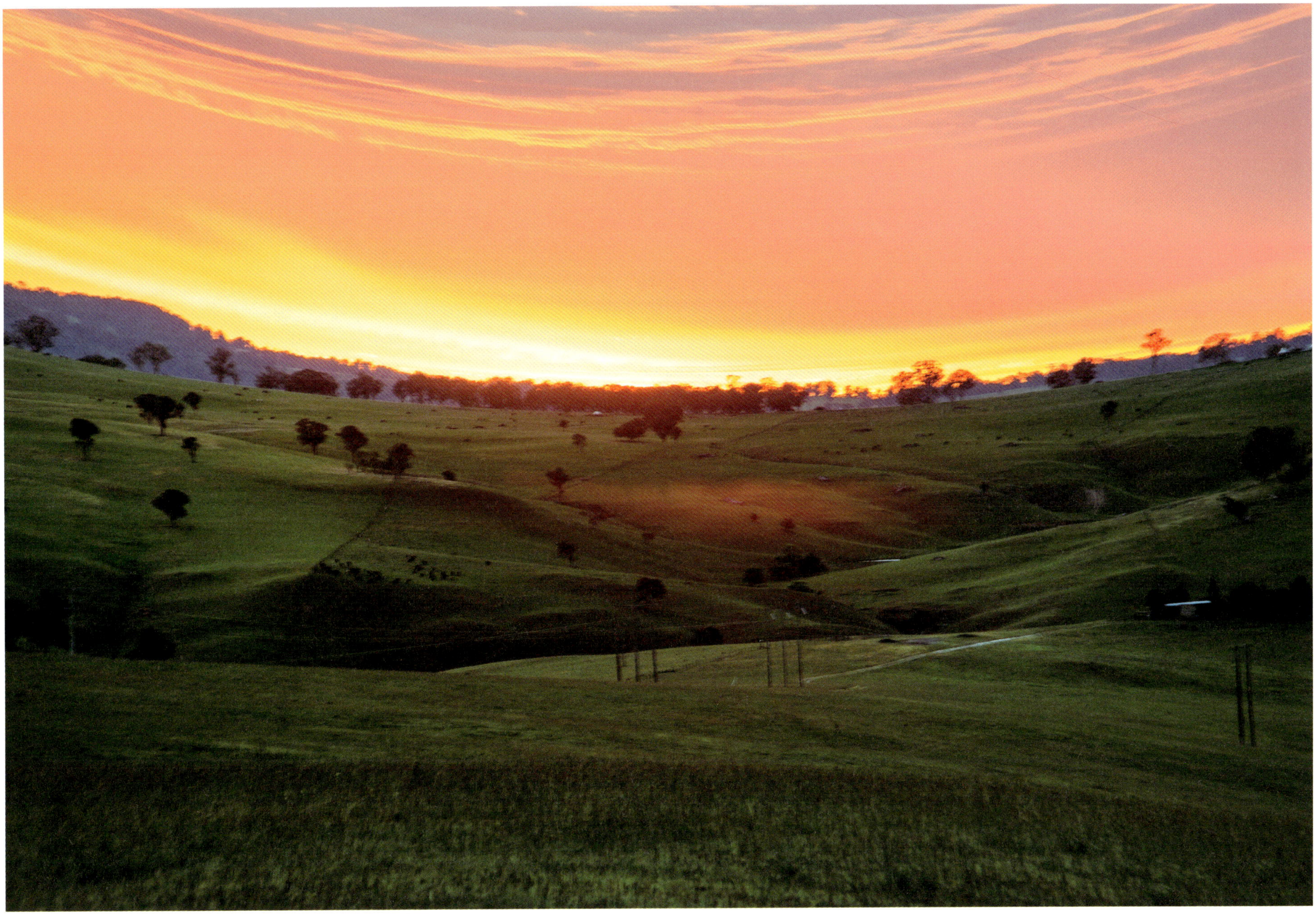

Loro Piana's people began to think about this story, imagining the journey of the merino sheep through such different lands, where man nonetheless succeeded in fulfilling its needs, rising to the challenge, looking for new solutions. In present-day growers in New Zealand and Australia Loro Piana found the same determination, the same way of tackling a challenge, the same passion for the very best that underpinned the ancient stories of explorations and discoveries. Centuries later, this beautiful sheep is still treated as a regal gift.

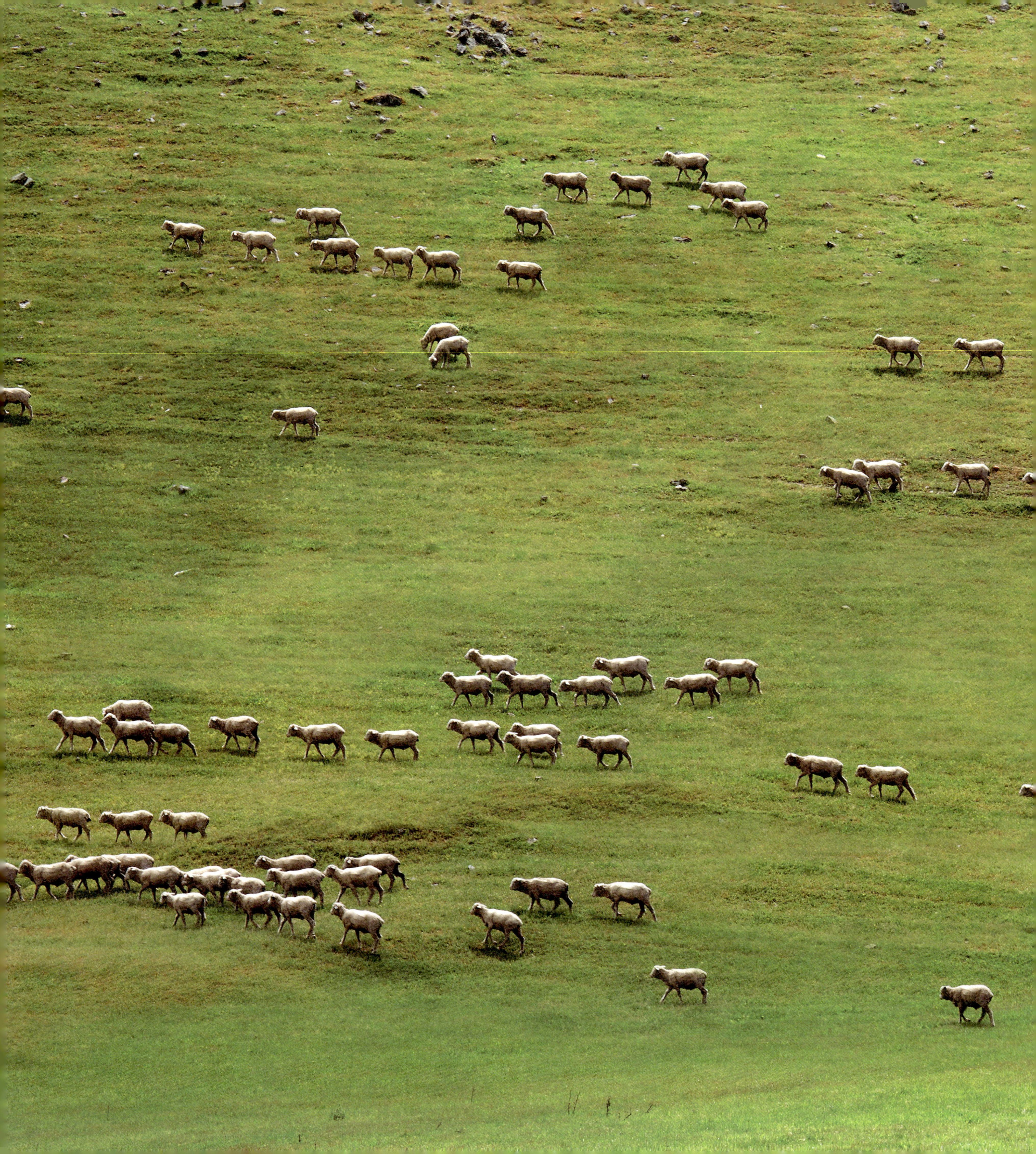

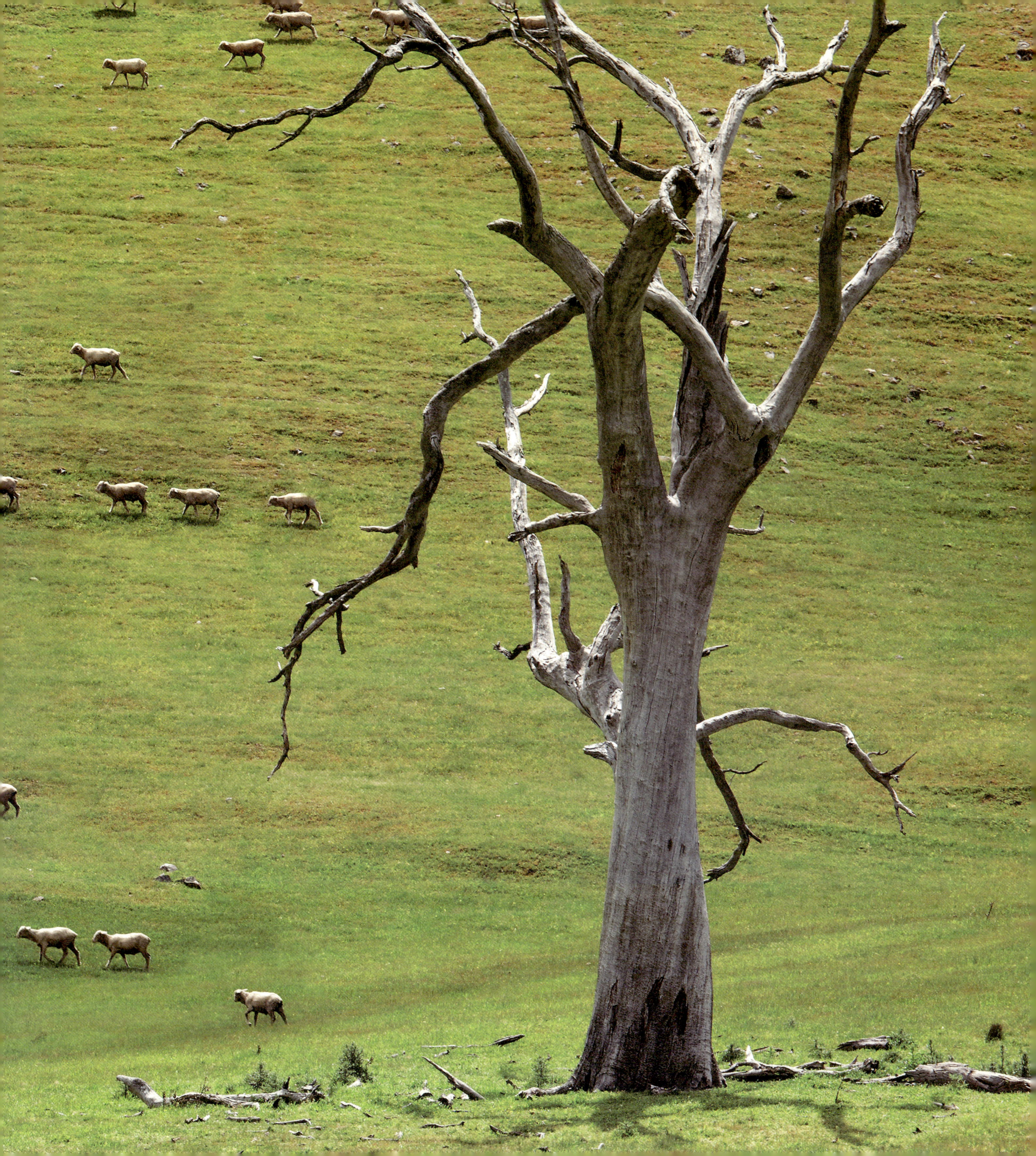

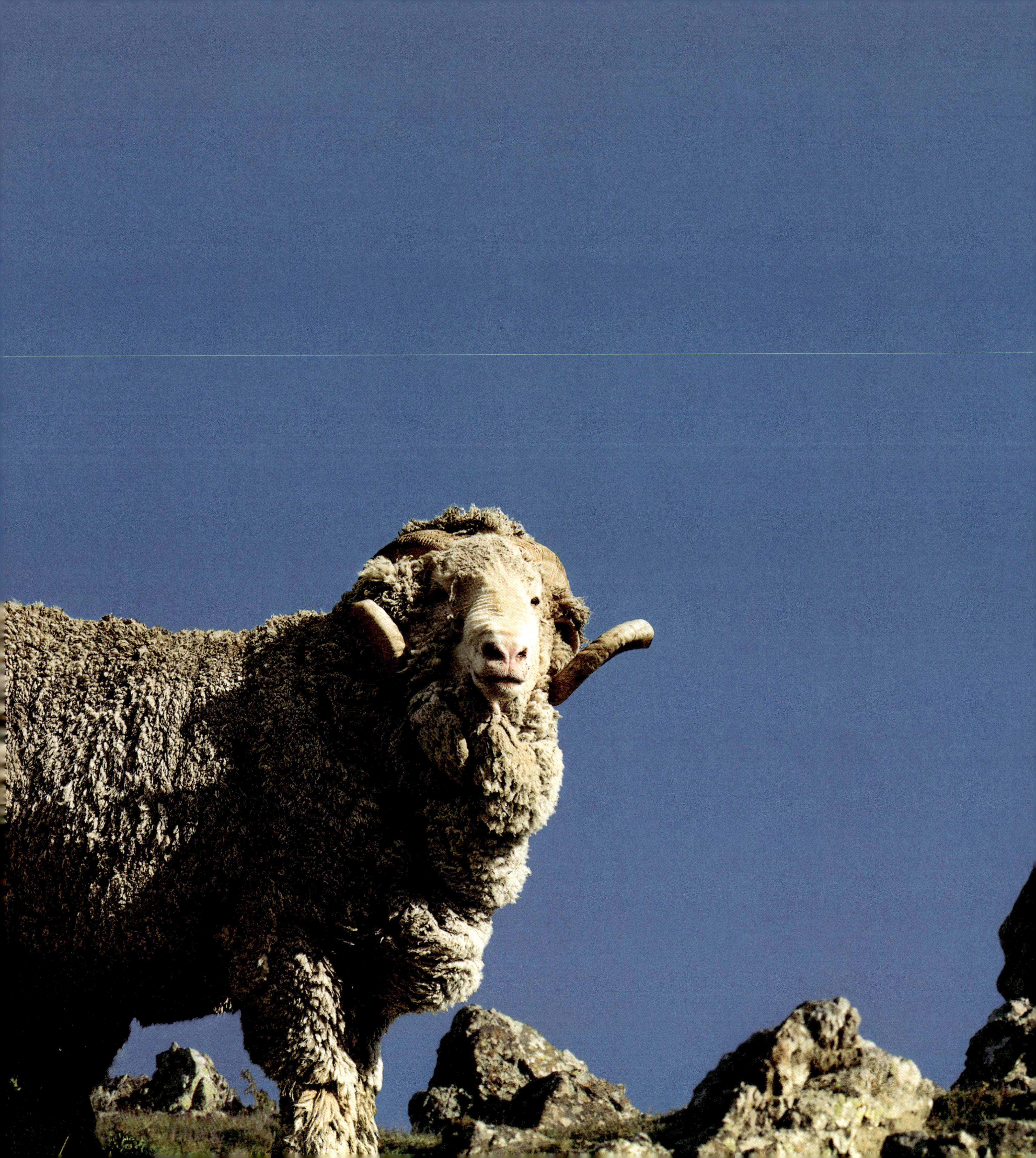

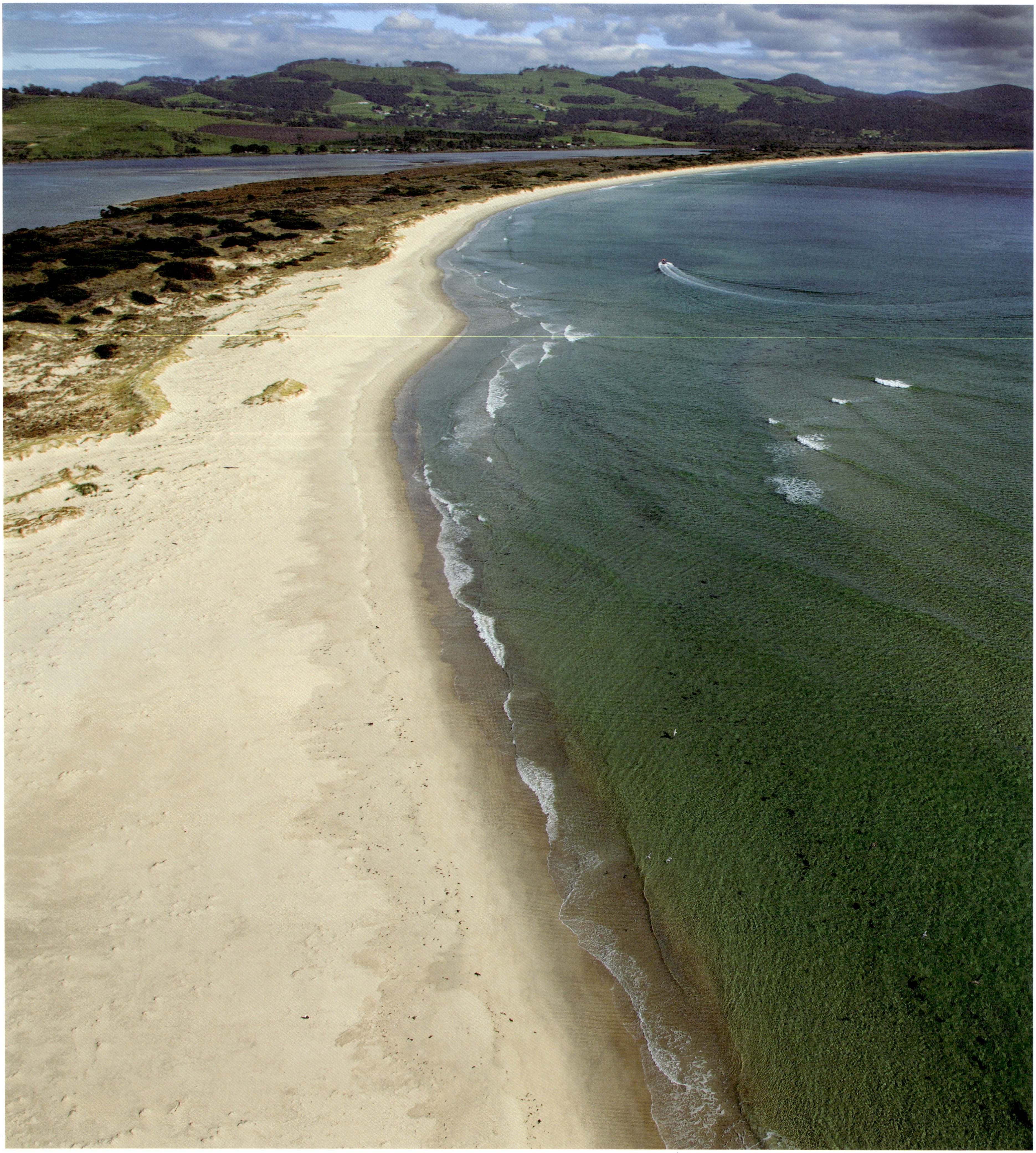

THE TASMANIAN COAST OFFERS A VARIETY OF STUNNING VIEWS: FROM LONG SANDY BEACHES THAT MEET THE SEA TO SHEER ROCKY CLIFFS THAT TOWER ABOVE THE WATER.

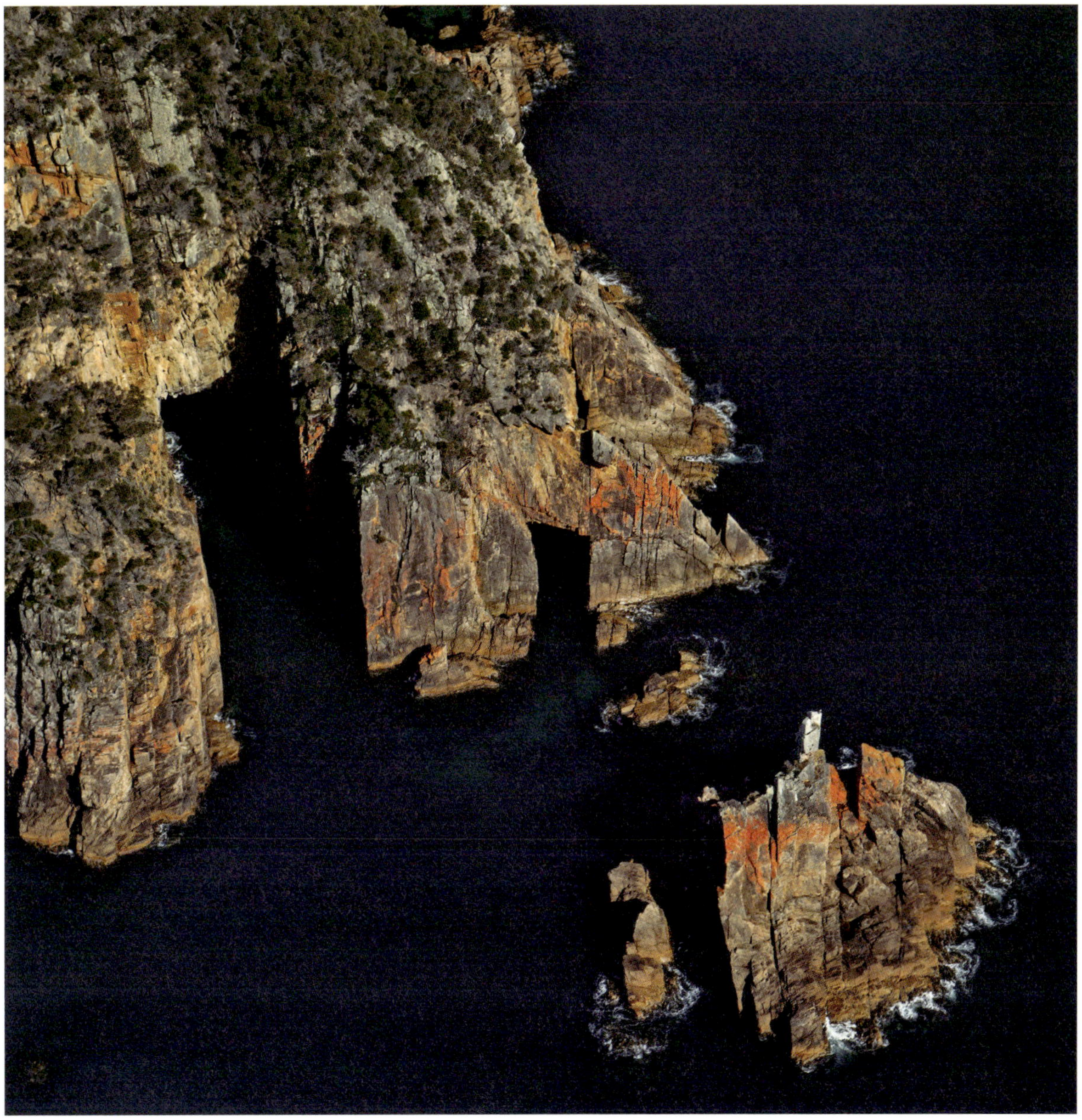

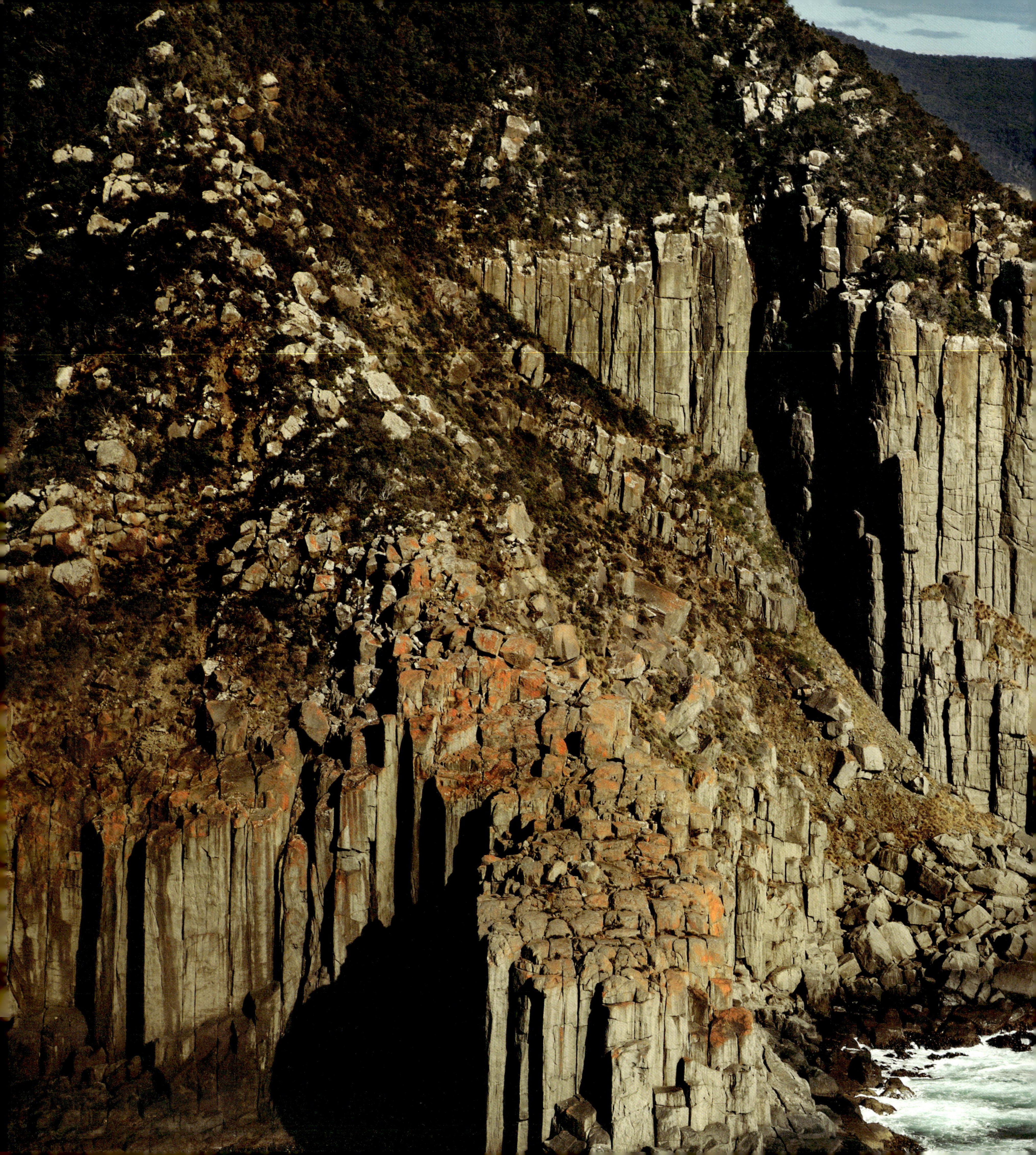

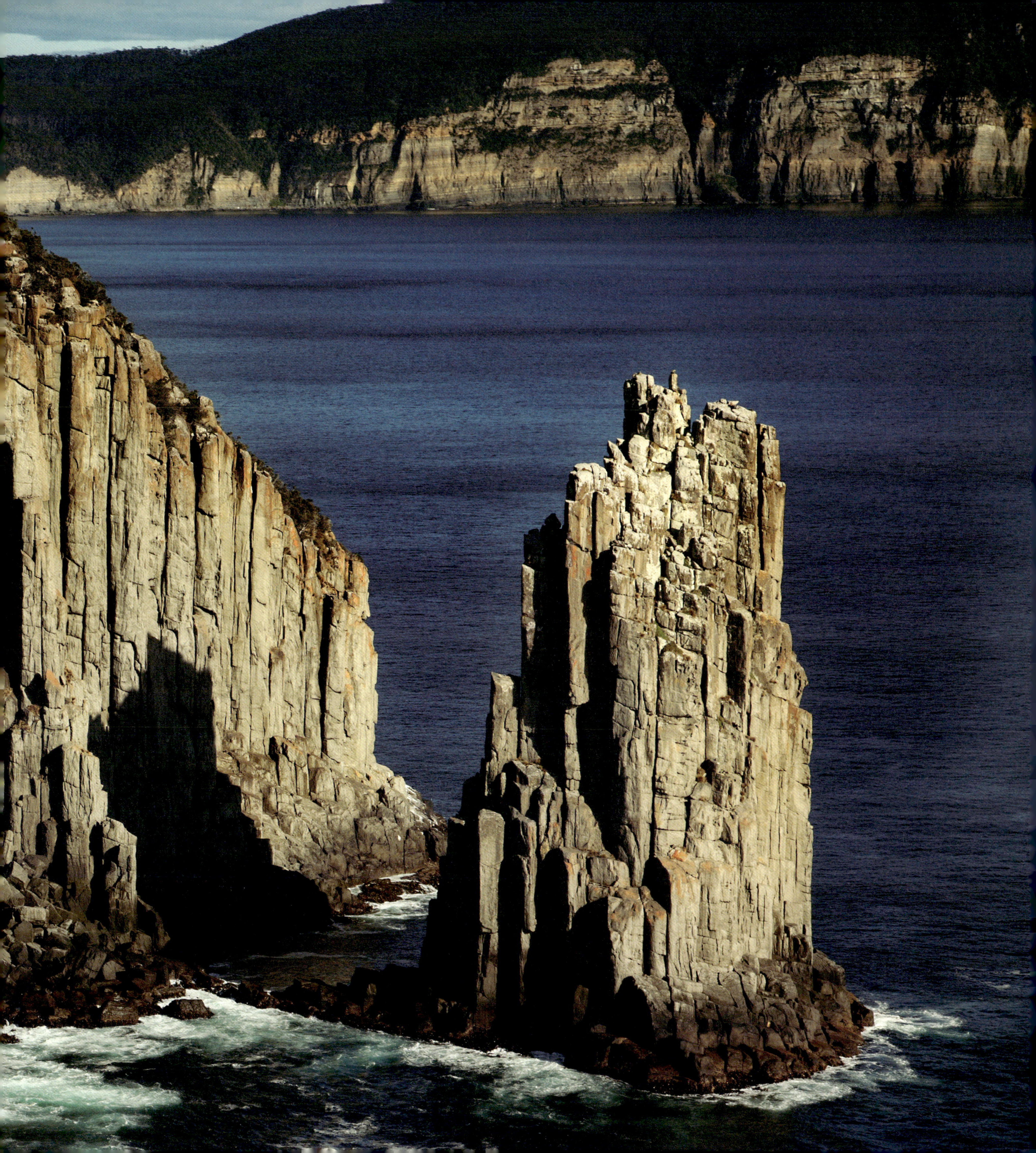

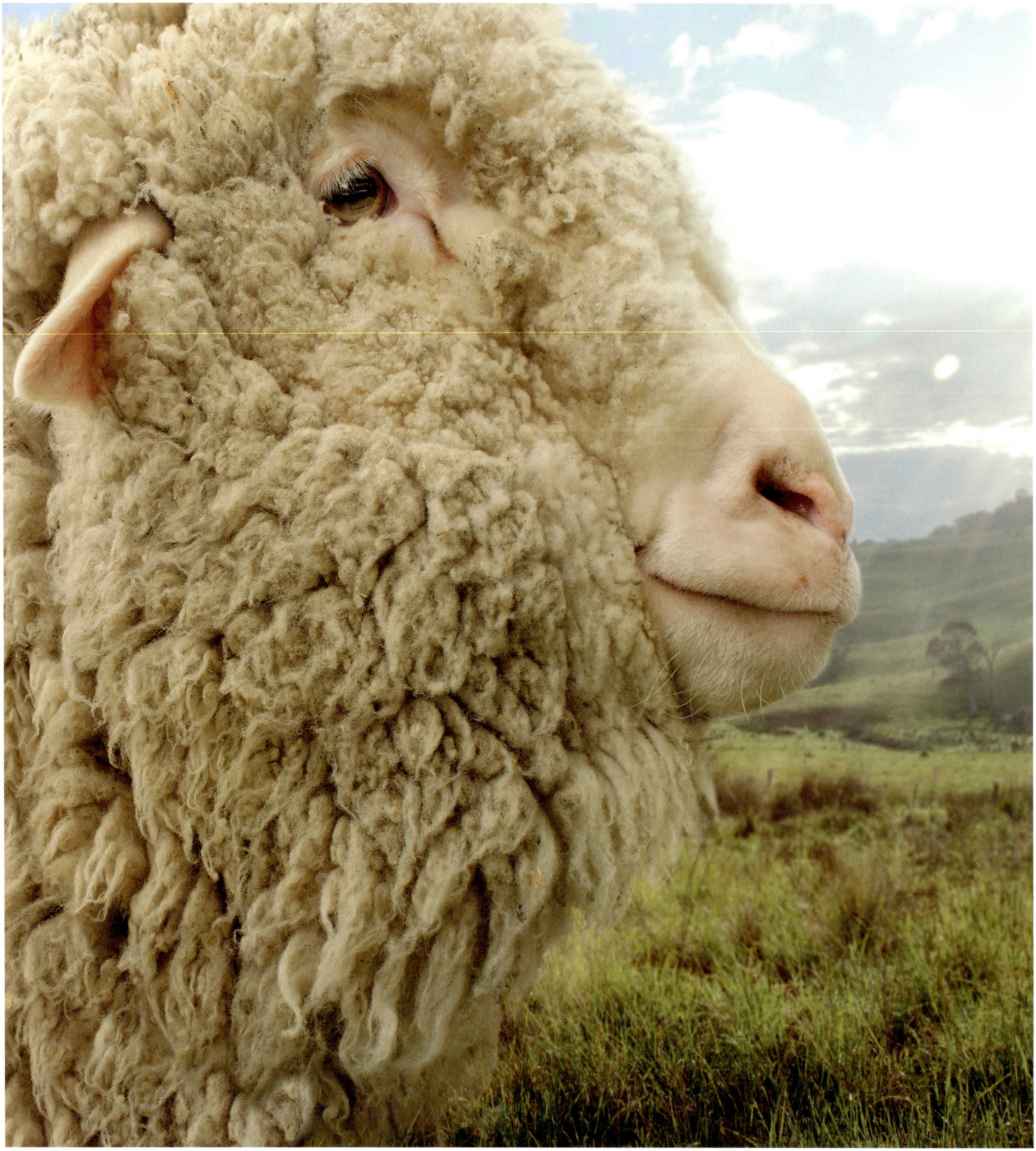

MERINO WOOL FIBRE HAS
A NATURALLY CRIMPED TEXTURE
THAT FORMS A PROTECTIVE BARRIER
AGAINST THE ANIMAL'S SKIN,
KEEPING IT PERFECTLY INSULATED.

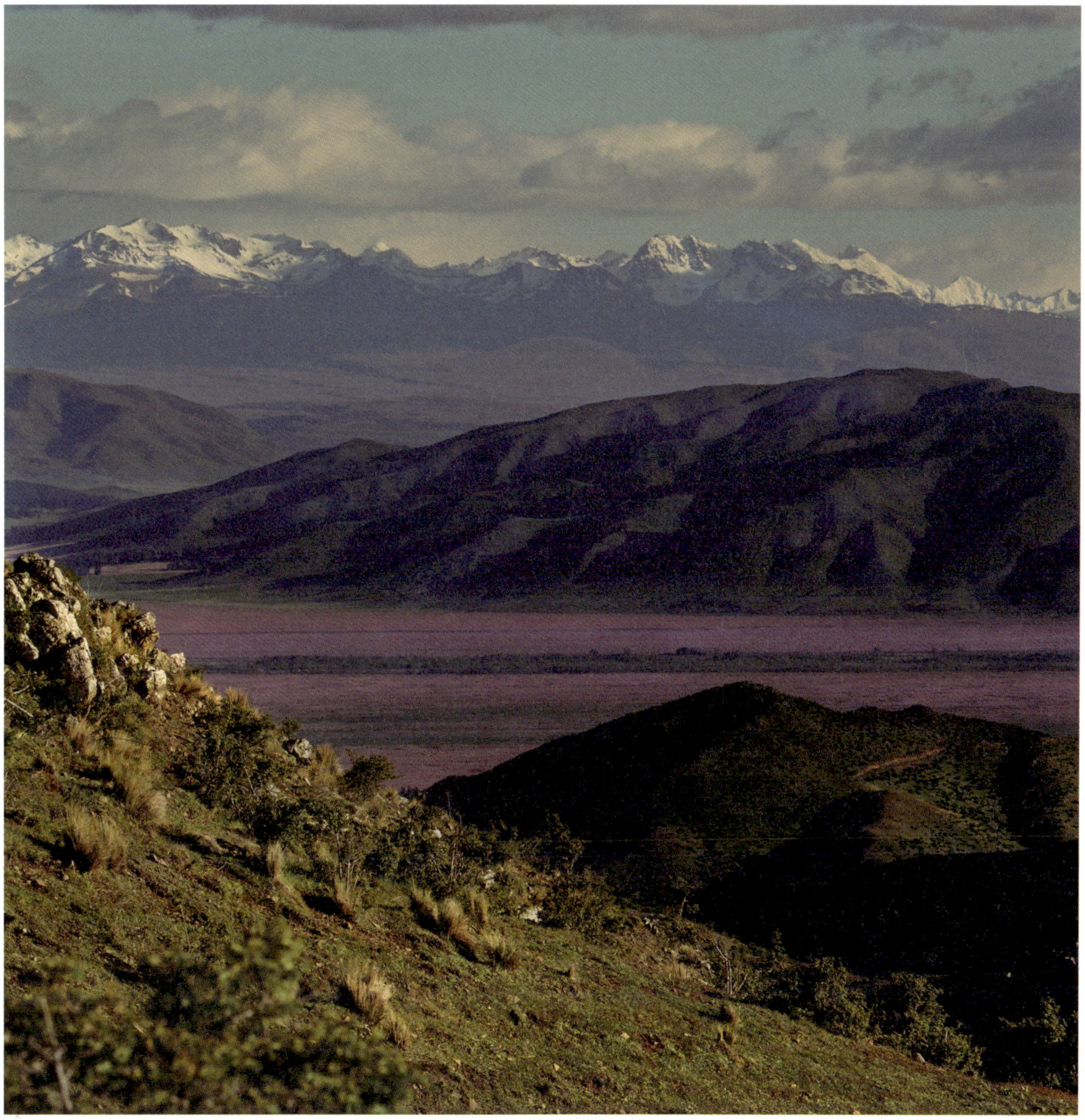

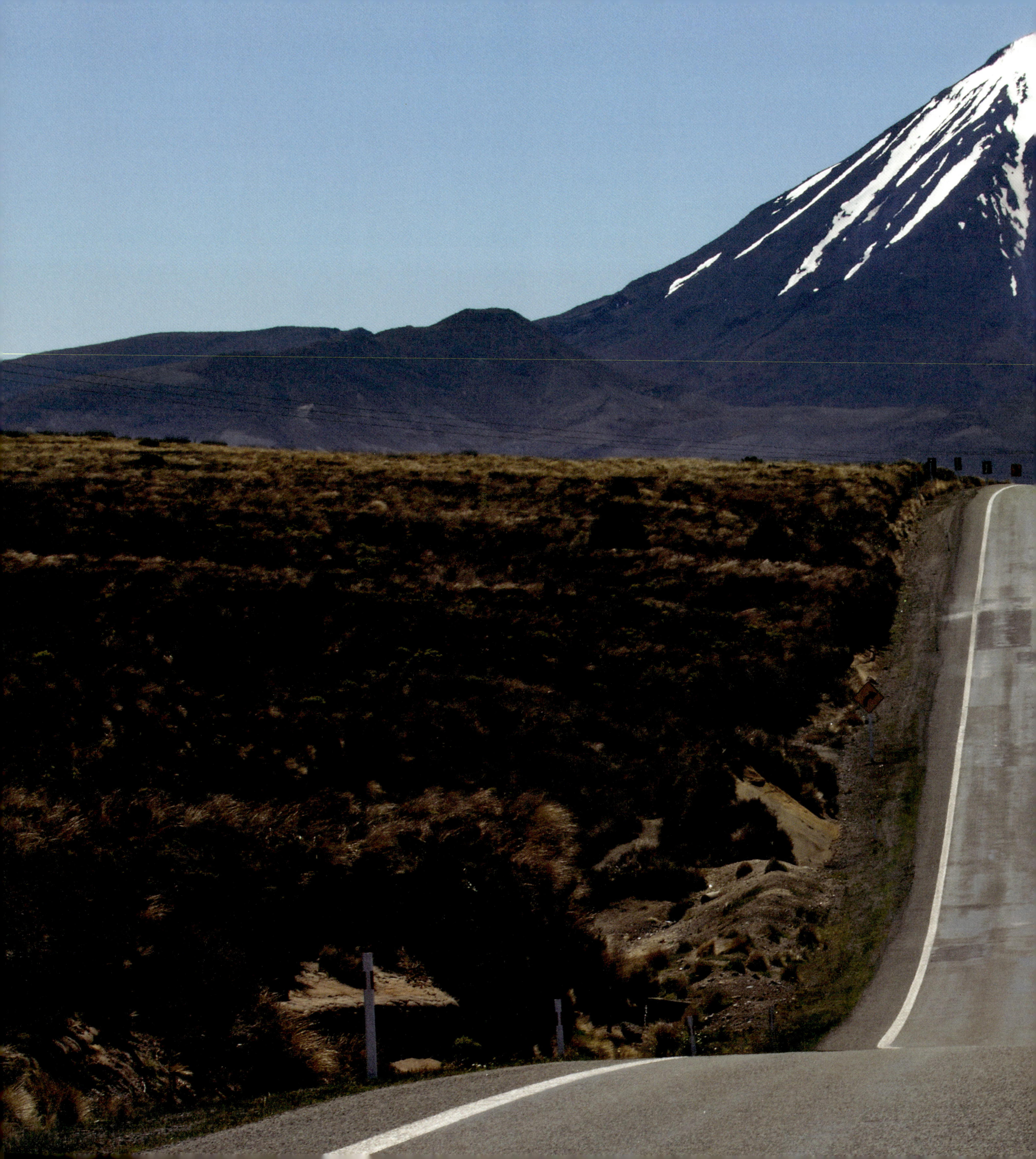

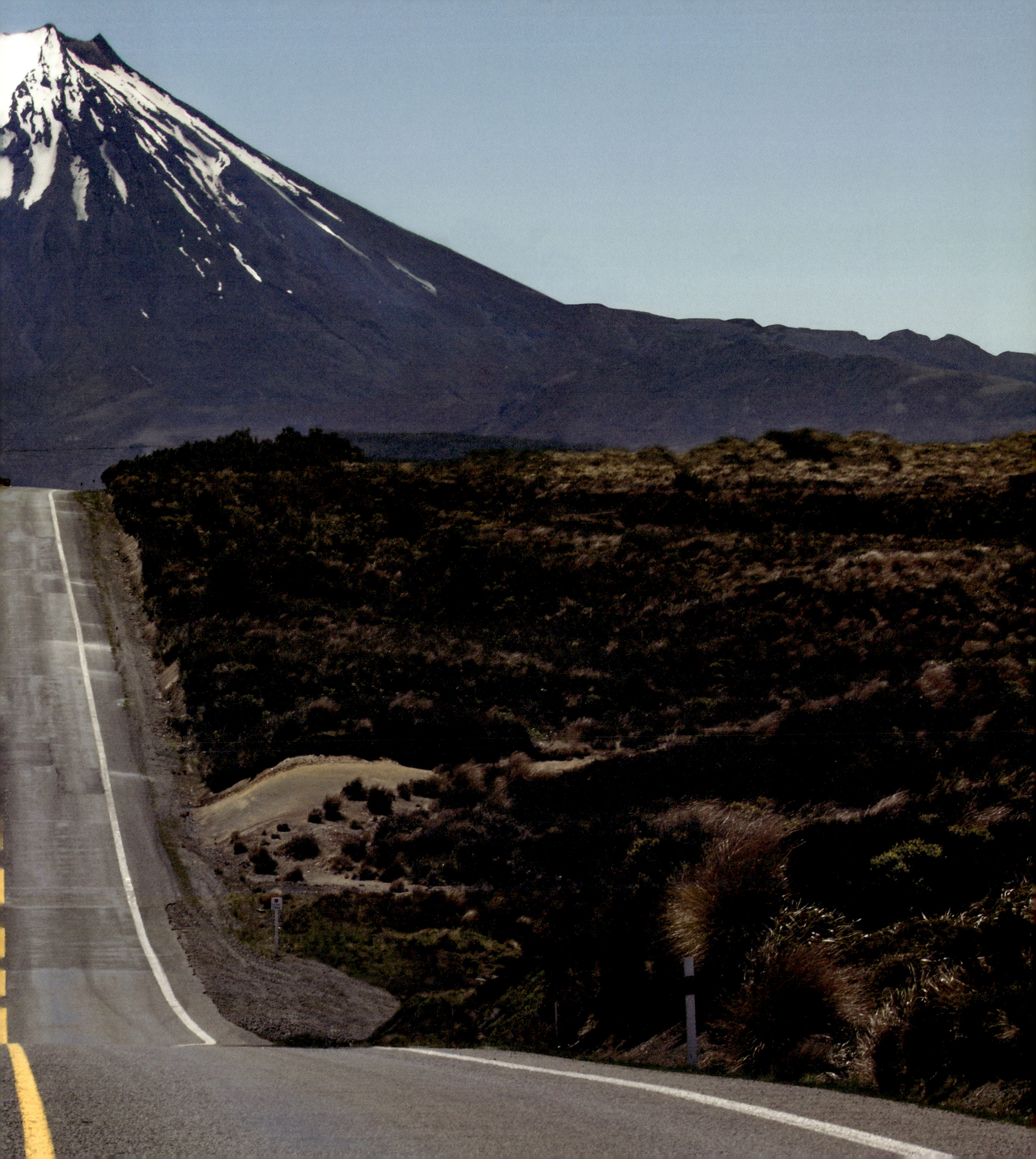

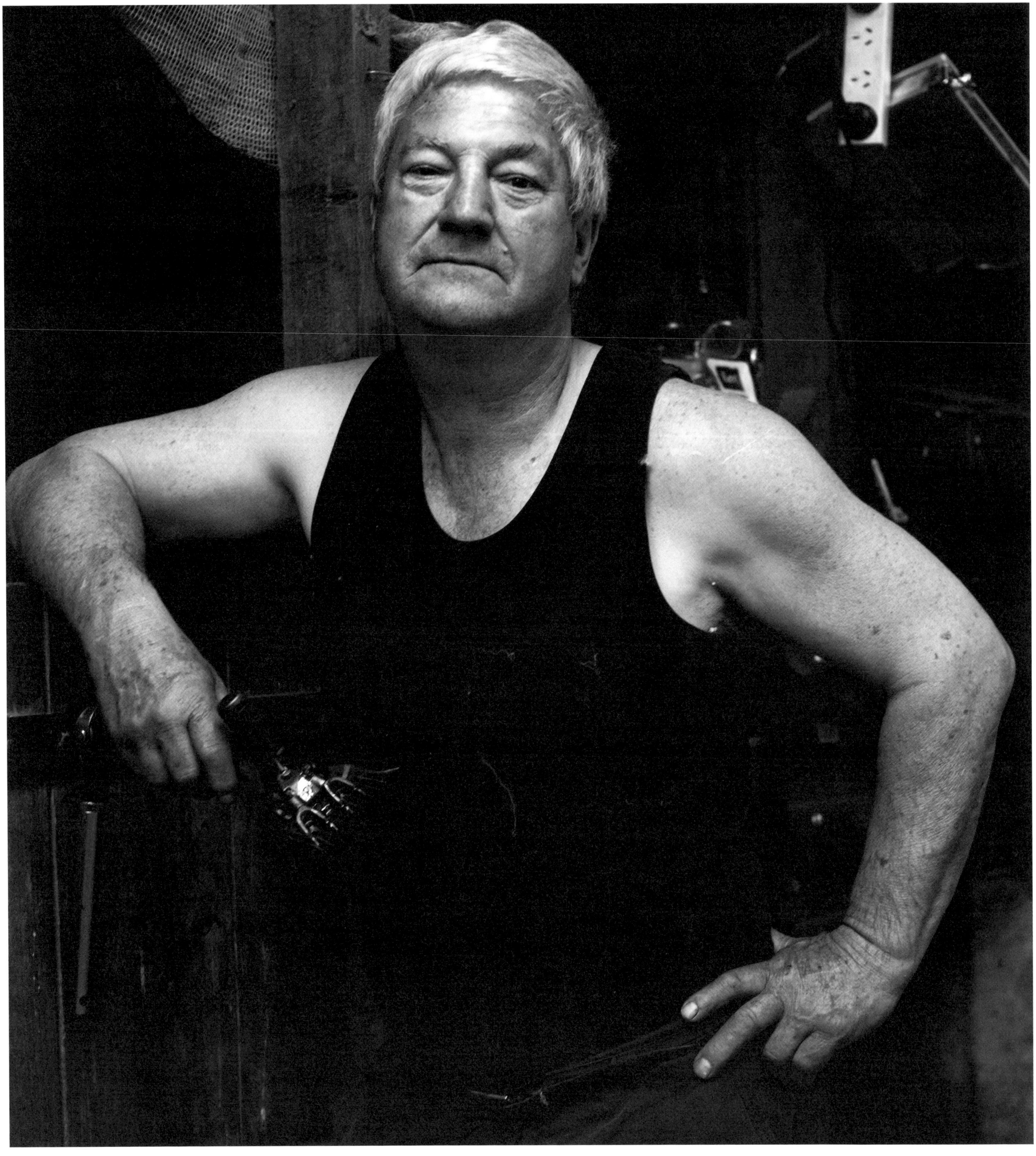

SEASONED WOOLGROWERS CAN MAKE A ROUGH ESTIMATE OF THE FINENESS OF A SHEEP'S FLEECE BY JUST LOOKING AT IT. THIS MIGHT SOUND LIKE A MAGICAL POWER, BUT ACCORDING TO THE GROWERS IT IS "JUST" EXPERIENCE.

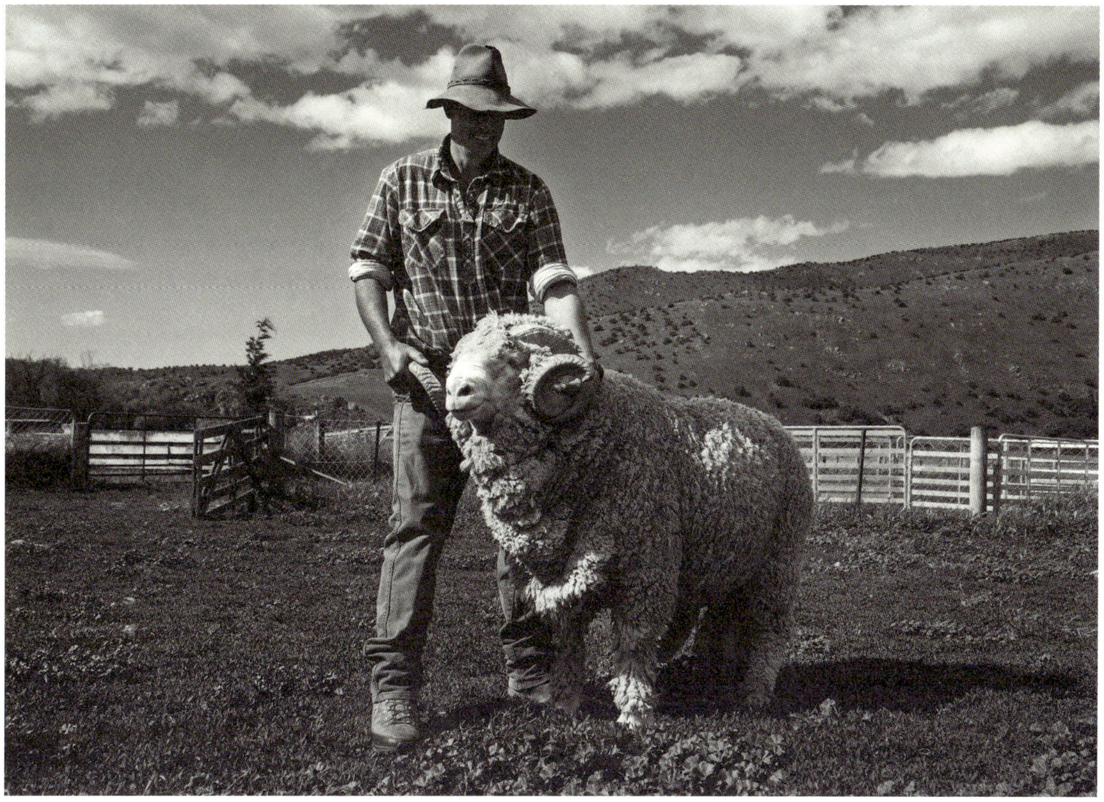

previous pages
The Desert Road is one of New Zealand's most beautiful scenic routes.

What for Loro Piana had originally started out as the need to source material for an exclusive product now became something broader, a concerted effort, a project: The Gift of Kings®. The Italian company began spending more and more time in the area, first and foremost to watch and learn. Finally they found the voice of the interlocutor they always sought in their ongoing dialogue with excellence: the voice of the farmers.

Each of them became an invaluable source of information, questions, potential paths. From the farmers Loro Piana learned that the merino sheep is both hardy and appreciative: if you treat these animals well, nourish them with love and method, protect and care for them, they will reward you with outstanding wool.

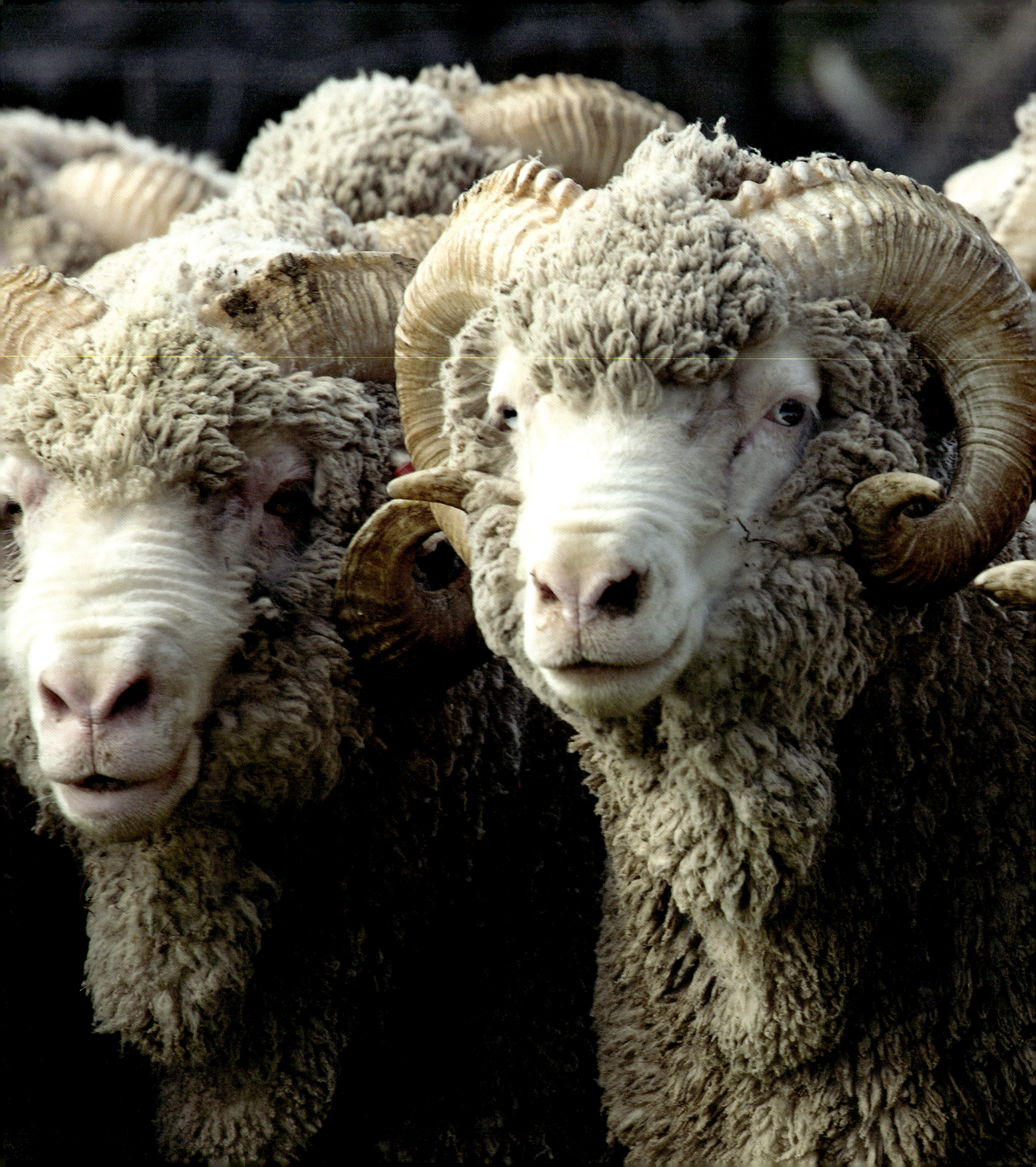

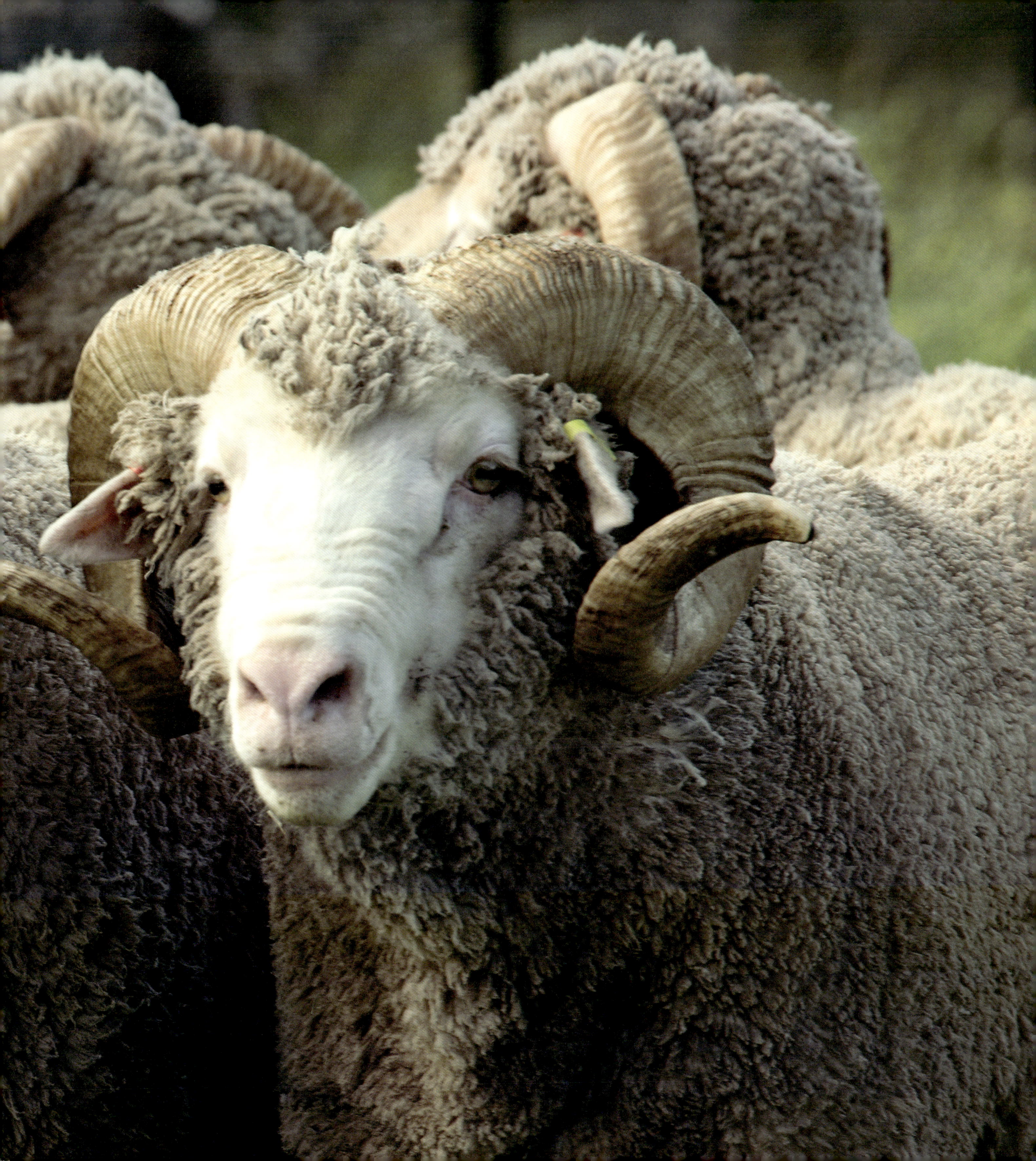

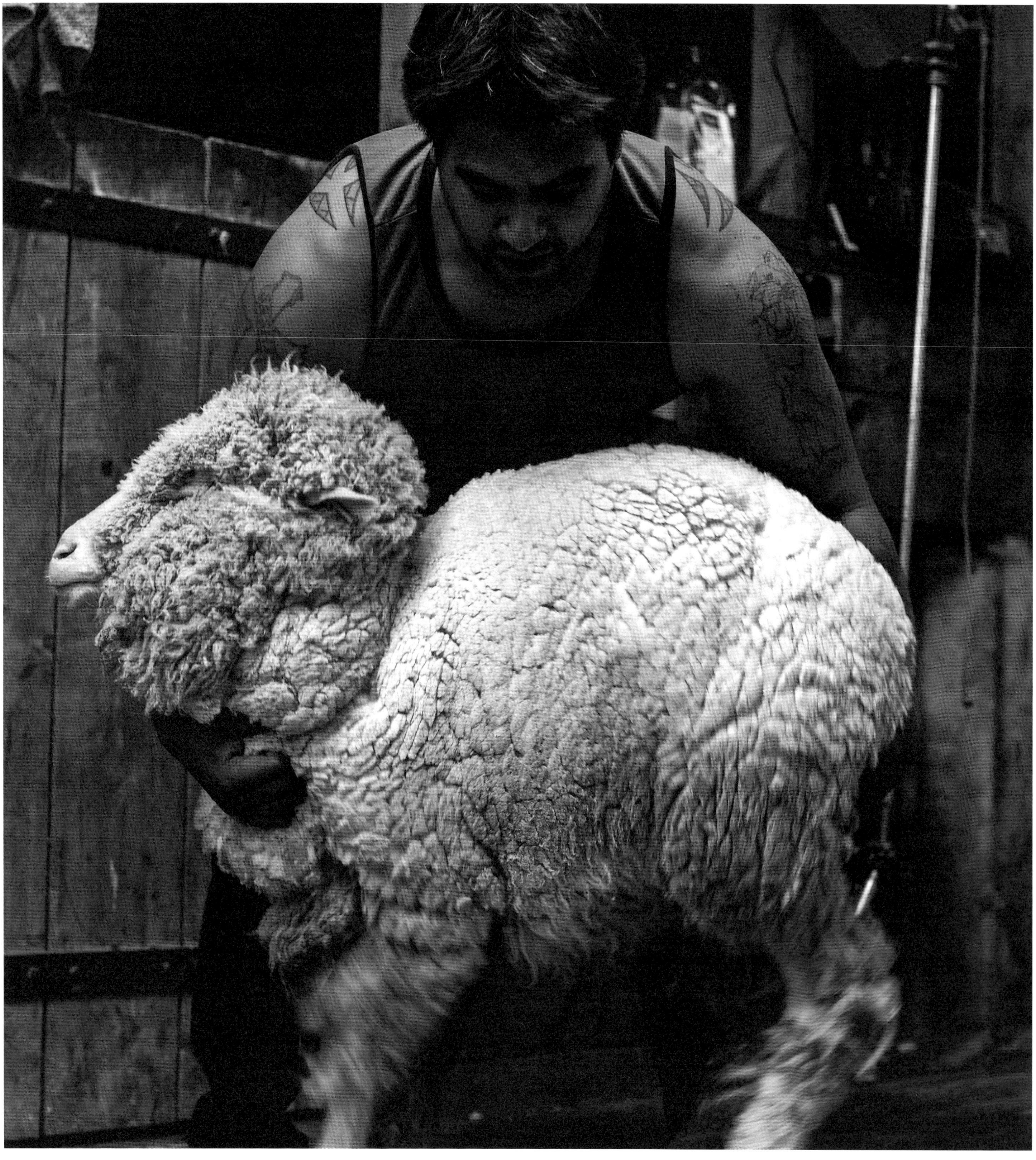

THE SHEEP OF NEW ZEALAND

AND THEIR PRECIOUS WOOL

IN THE HANDS OF THE GUARDIANS

OF THIS GIFT OF KINGS:

THE PAYNE FAMILY OF VISULEA.

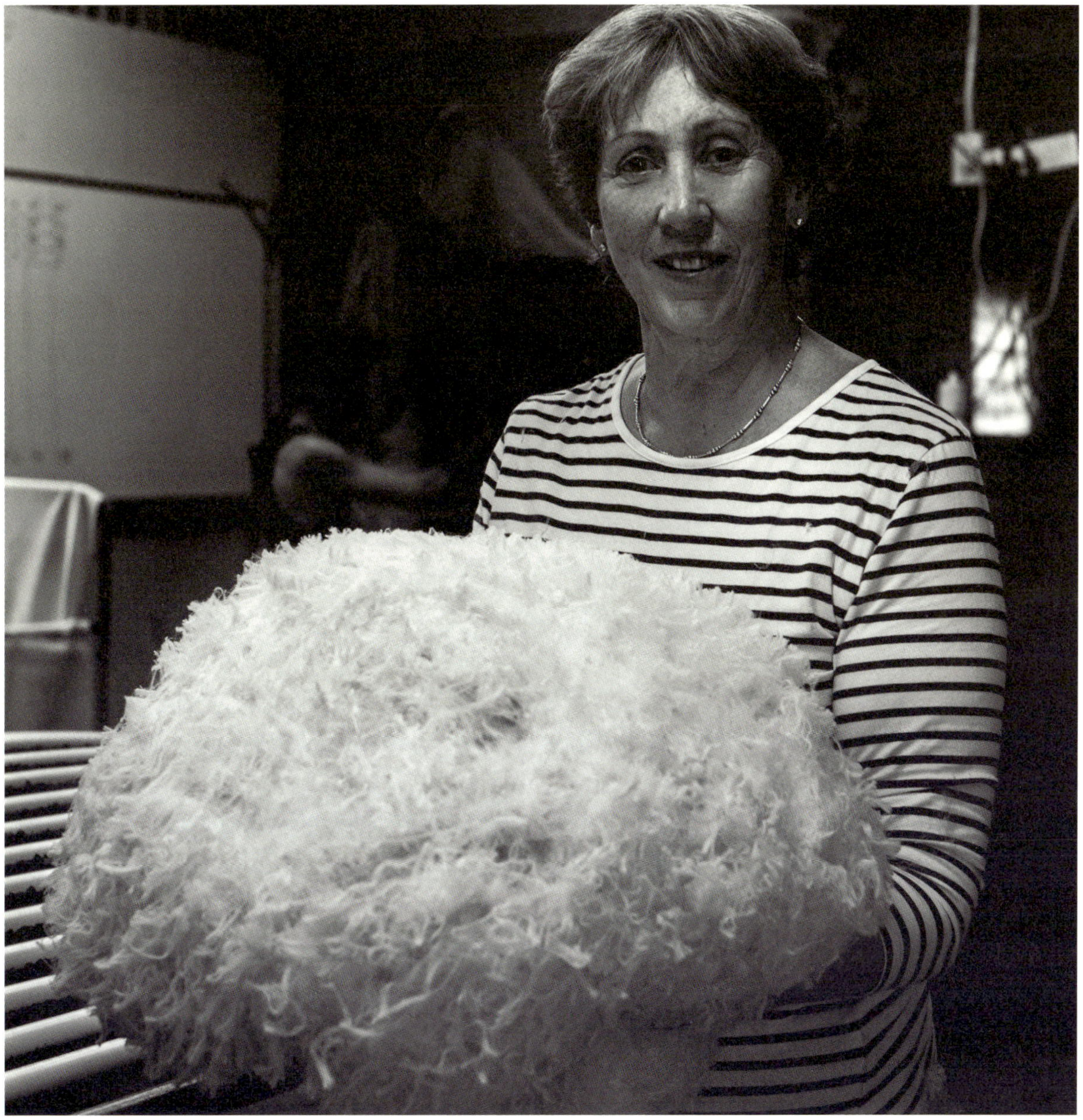

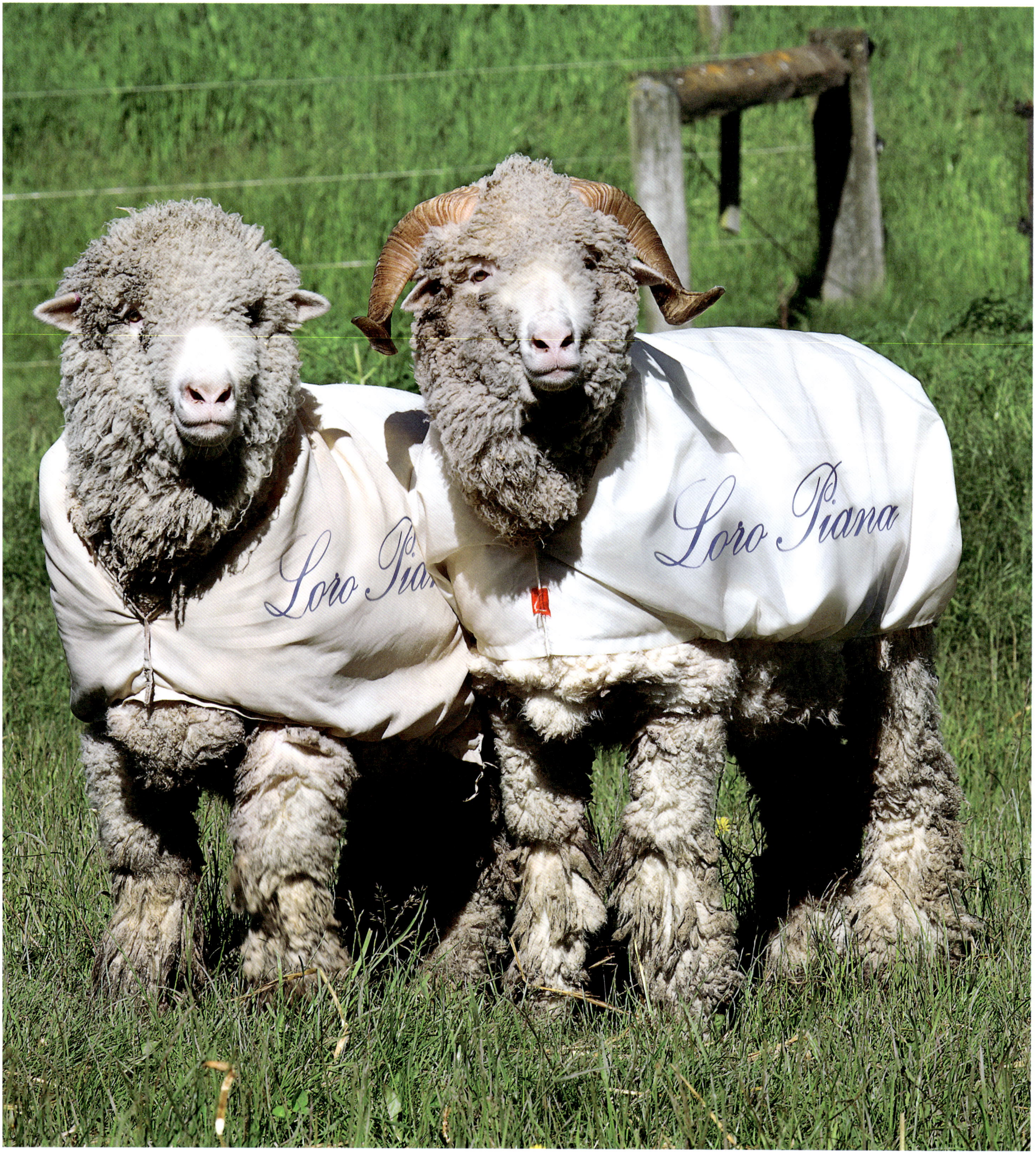
Loro Piana
Loro Piana

SPECIAL COVERS PROTECT THE FLEECES OF THE SHEEP ON THE PAYNE FARM IN VISULEA, NEW ZEALAND.

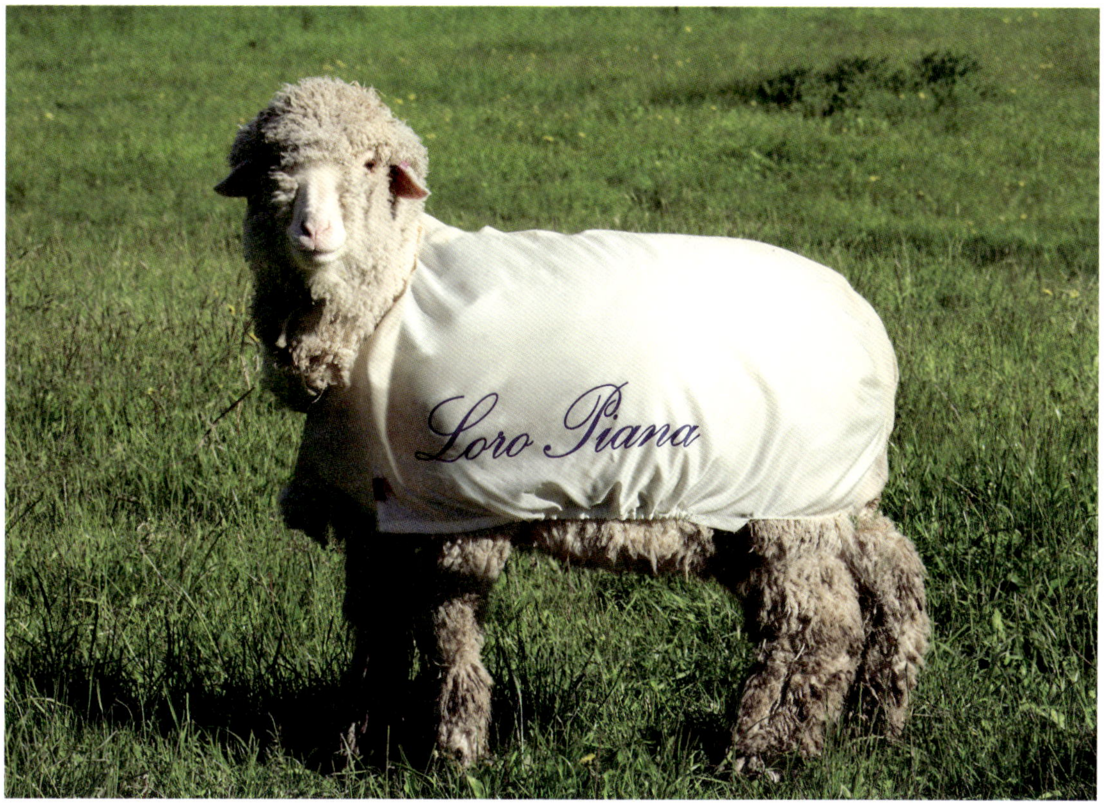

Coats specially designed to protect the sheep from the elements, frequently moving the flocks to find suitable grazing land, and regular monitoring of their health, appearance and behaviour: these are just a few of the aspects that all those working on a sheep farm have to apply with method and constancy. This is the way forward for naturally healthy sheep that produce increasingly fine, superlative quality wool. Donald Burnett was living proof of it: the wool he showed to Loro Piana in 1997 was incredibly fine, only 13.8 microns in diameter.

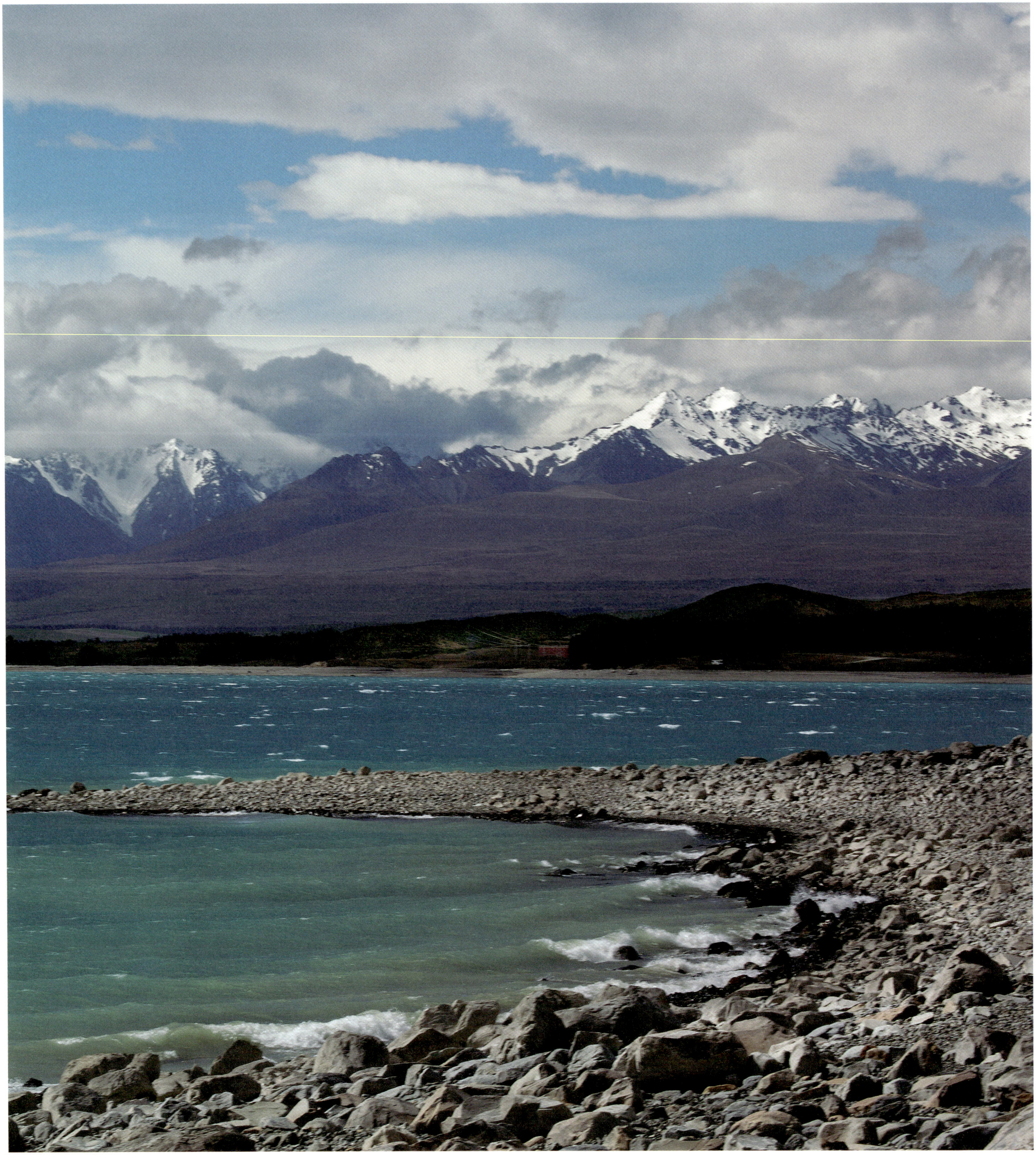

WATER IS A KEY PART

OF NEW ZEALAND'S LANDSCAPE:

LAKE WANAKA IS ONE OF

THE COUNTRY'S LARGEST.

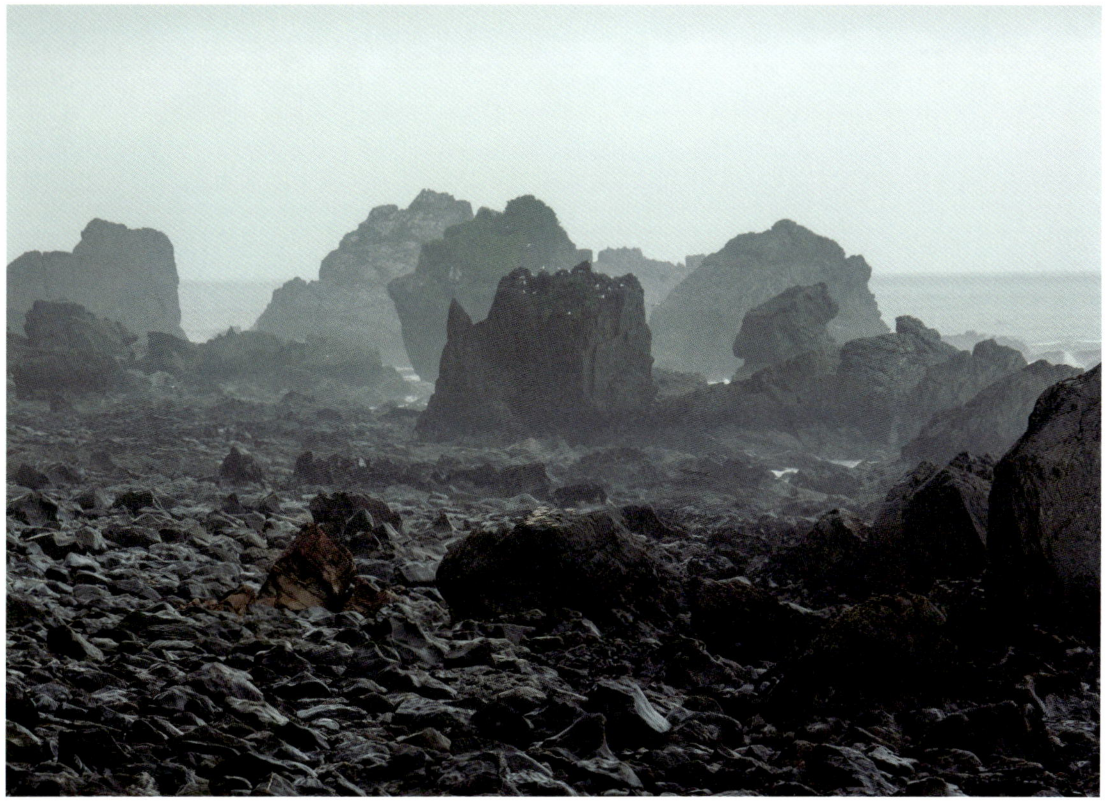

Barrytown Coast, New Zealand.

following pages
The Mount Cook Valley in New Zealand,
home to the highest mountains
in the country.

The vision of this man, and all of Loro Piana's woolgrowers, contains the poetry of an age-old craft, the lessons of the past and the broad horizons of a window open on the future. When this project got under way, buzz words like environmental protection, sustainability and zero footprint were only just taking off, but in these parts they were already a reality.

Enthusiasm for the project reached Quarona, Loro Piana's headquarters in Italy, where work began on trialling methods suited to turning this very special raw material into yarn and then fabric, with the utmost respect for its unique properties. Increasingly refined and effective technology was developed, guided by the principle of making the most of this natural product.

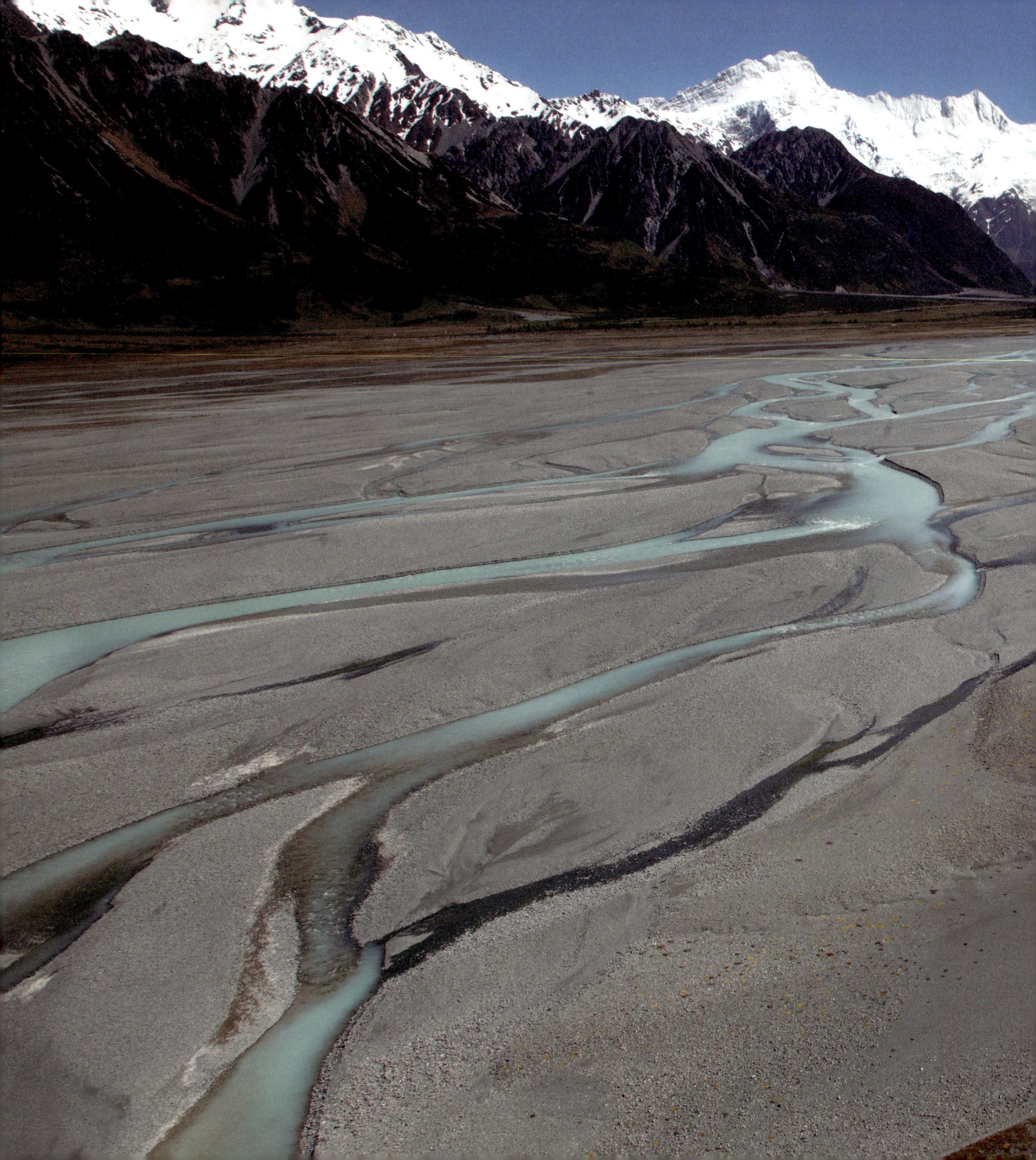

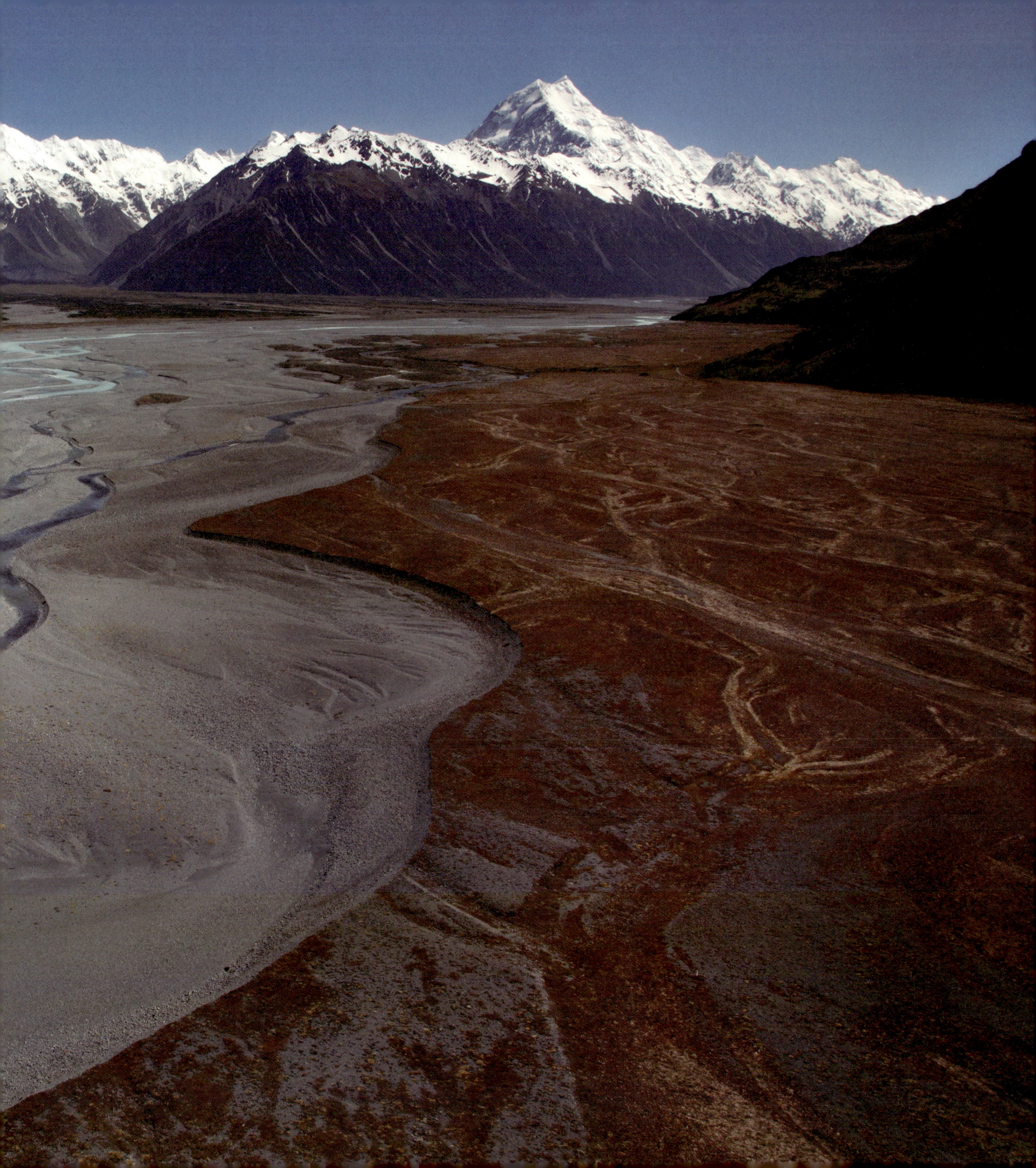

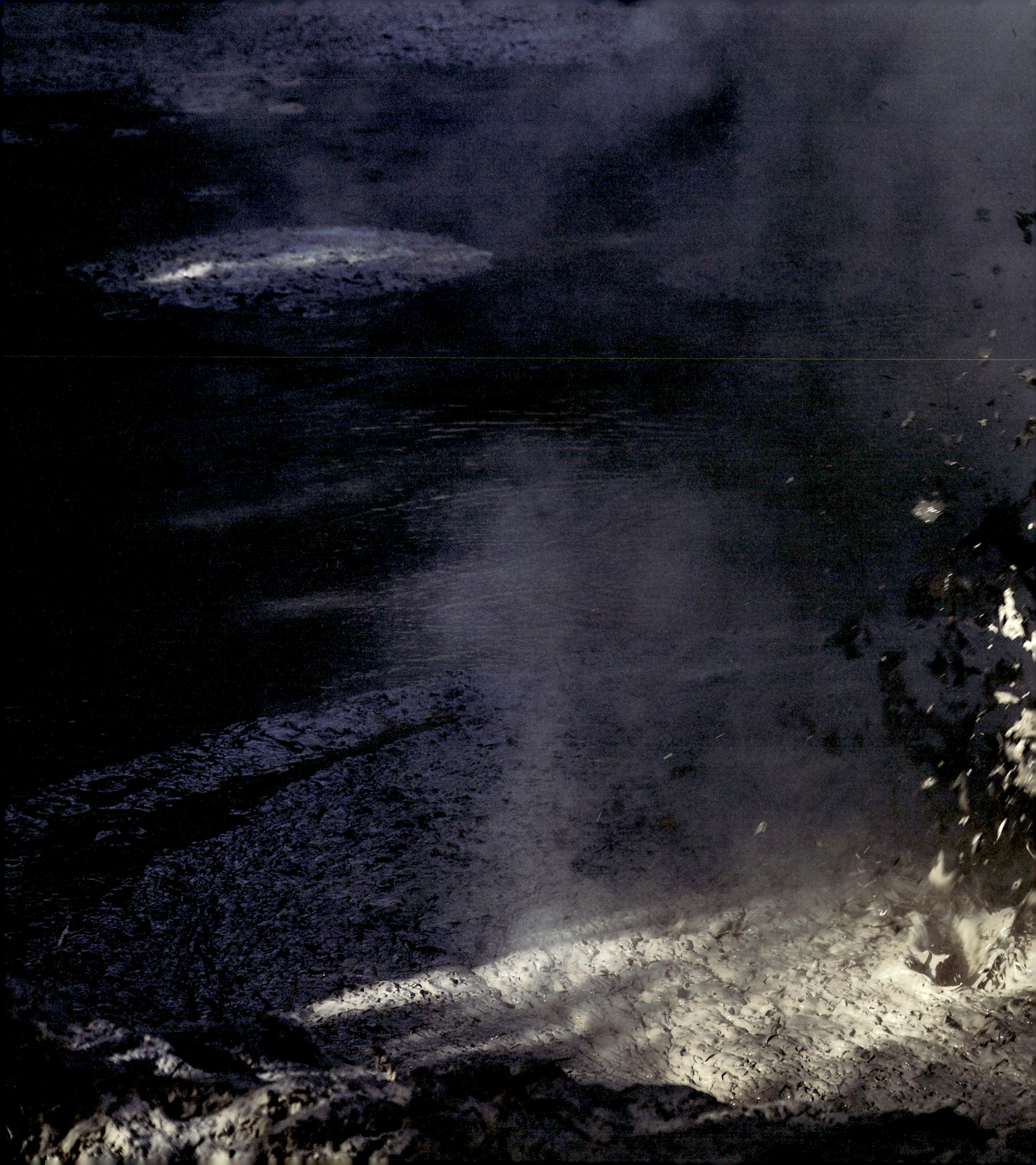

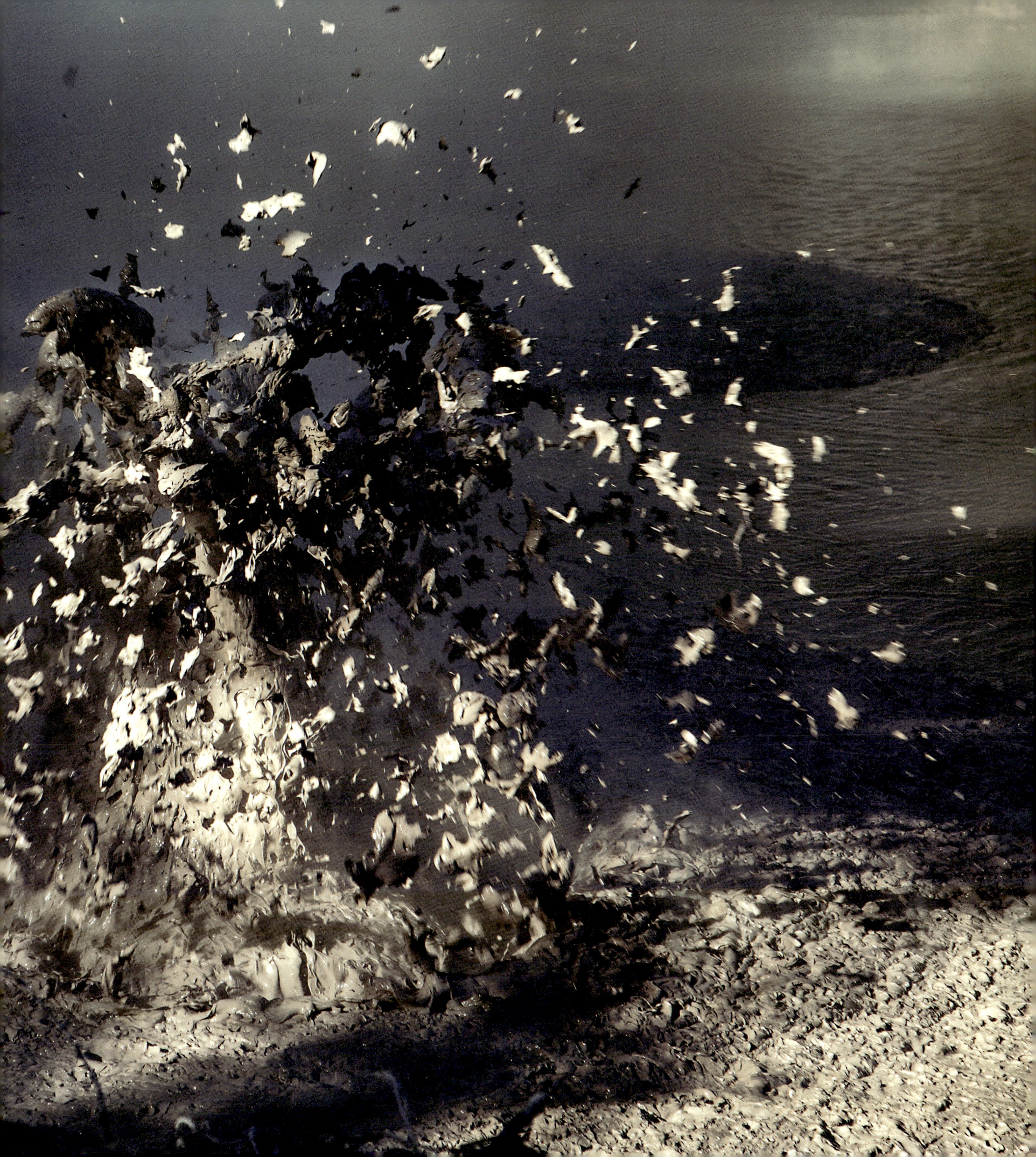

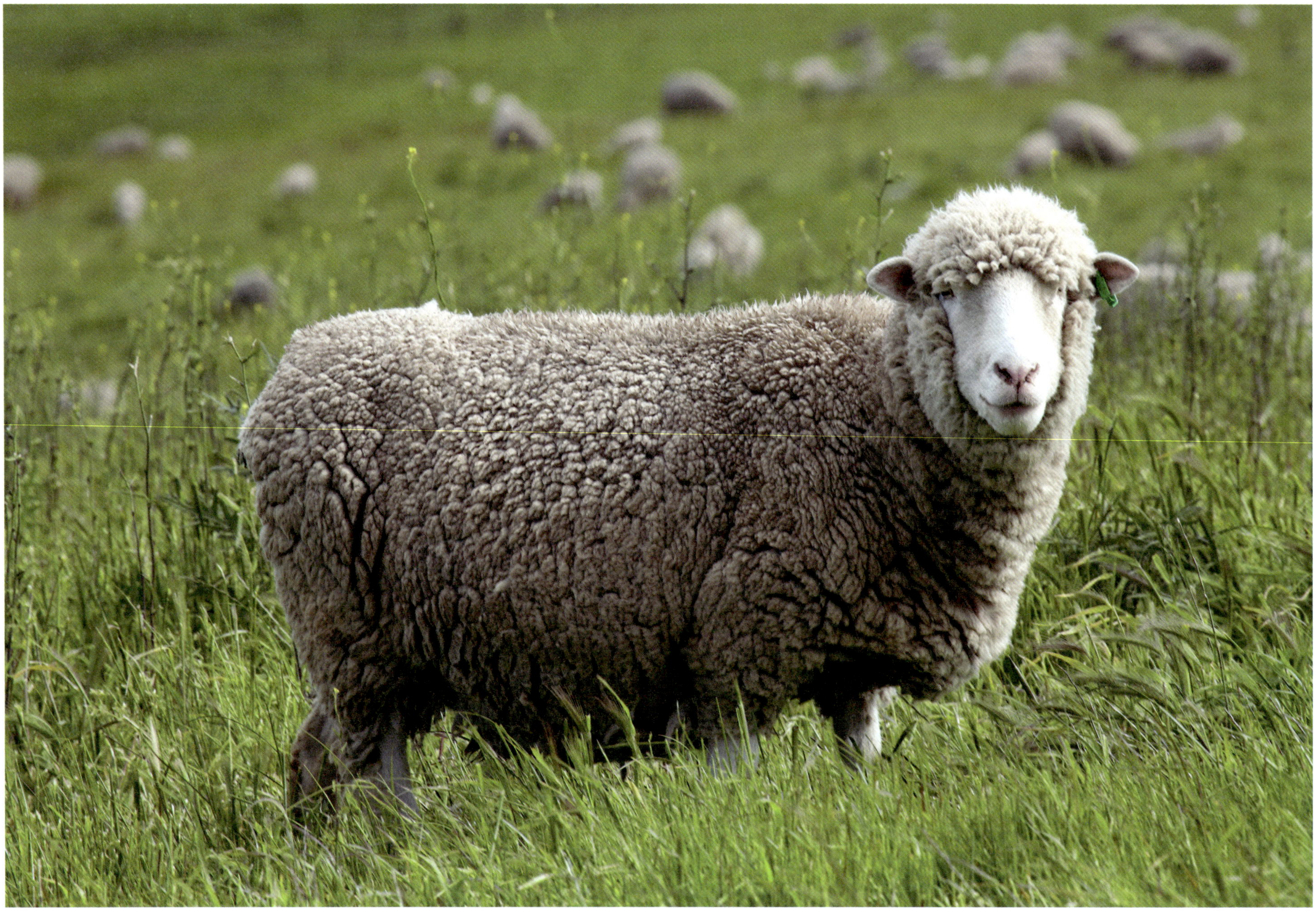

previous pages
The Rotorua area, which hosts the lake of the same name, boasts geysers that are among New Zealand's most spectacular geothermal wonders.

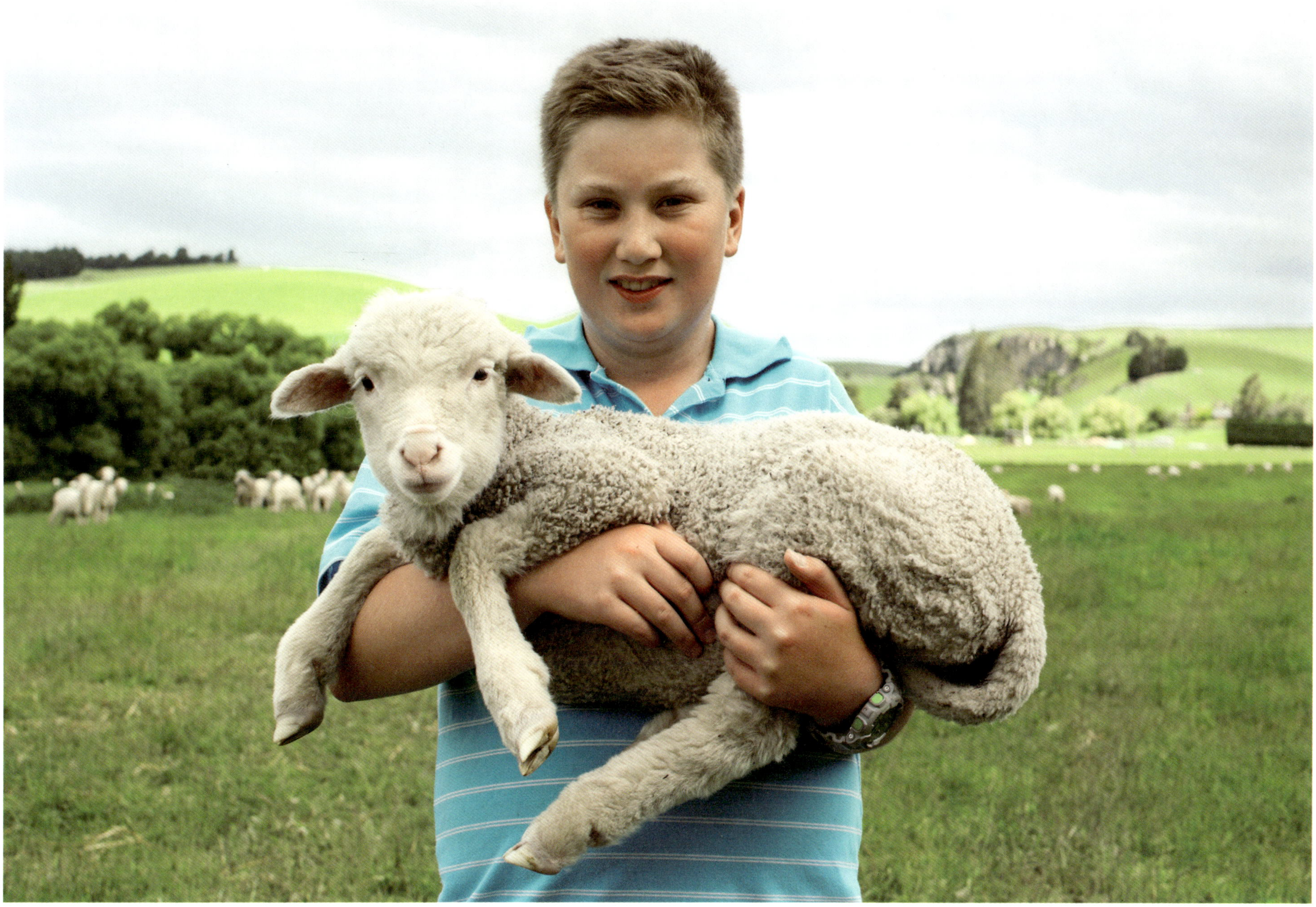

On the farms nutrition is all important.
By rotating grazing land and applying holistic principles the growers guarantee that each sheep's dietary requirements are being met. Only healthy animals produce fleece fit for a king.

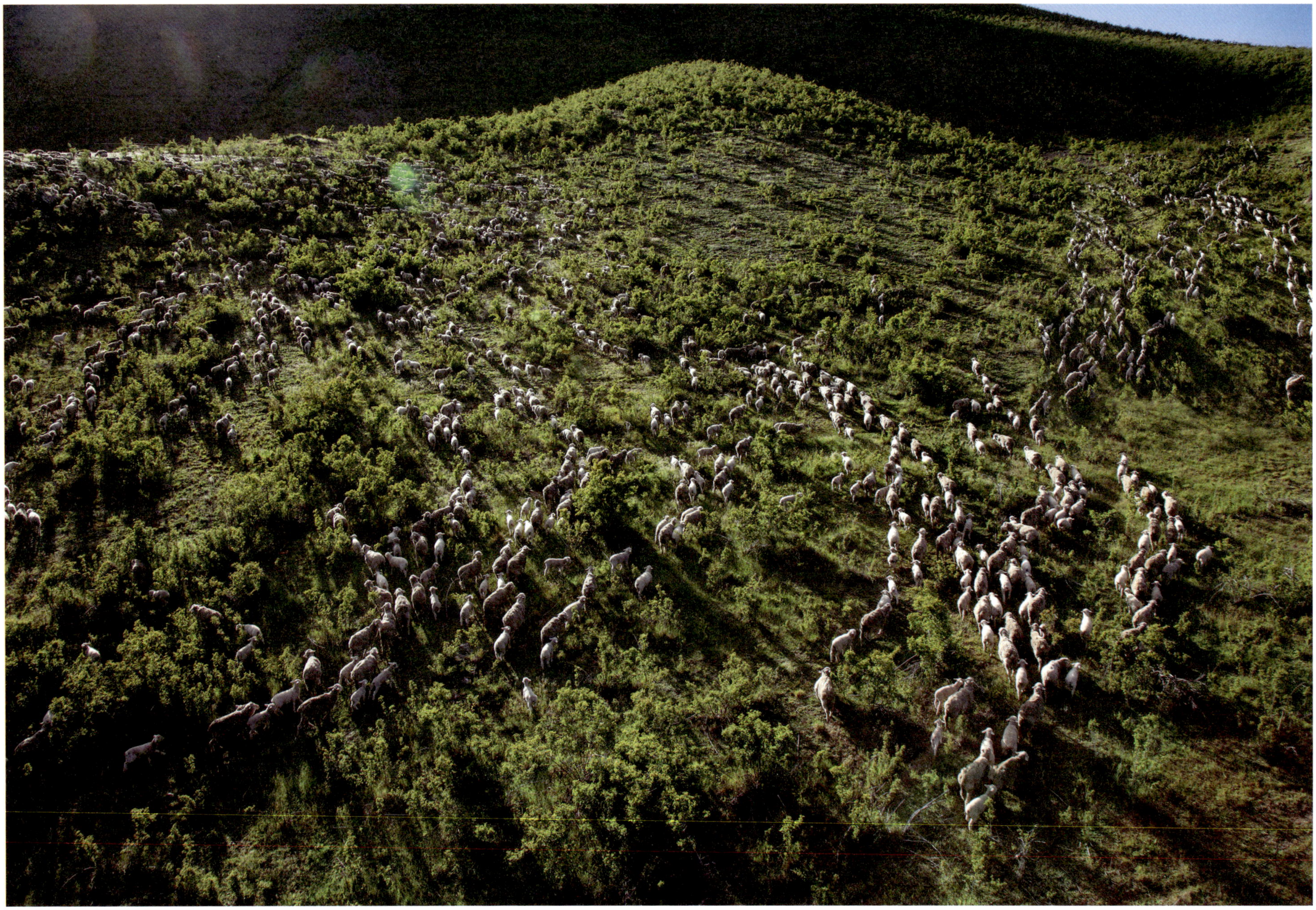

following pages

The West Coast, New Zealand.

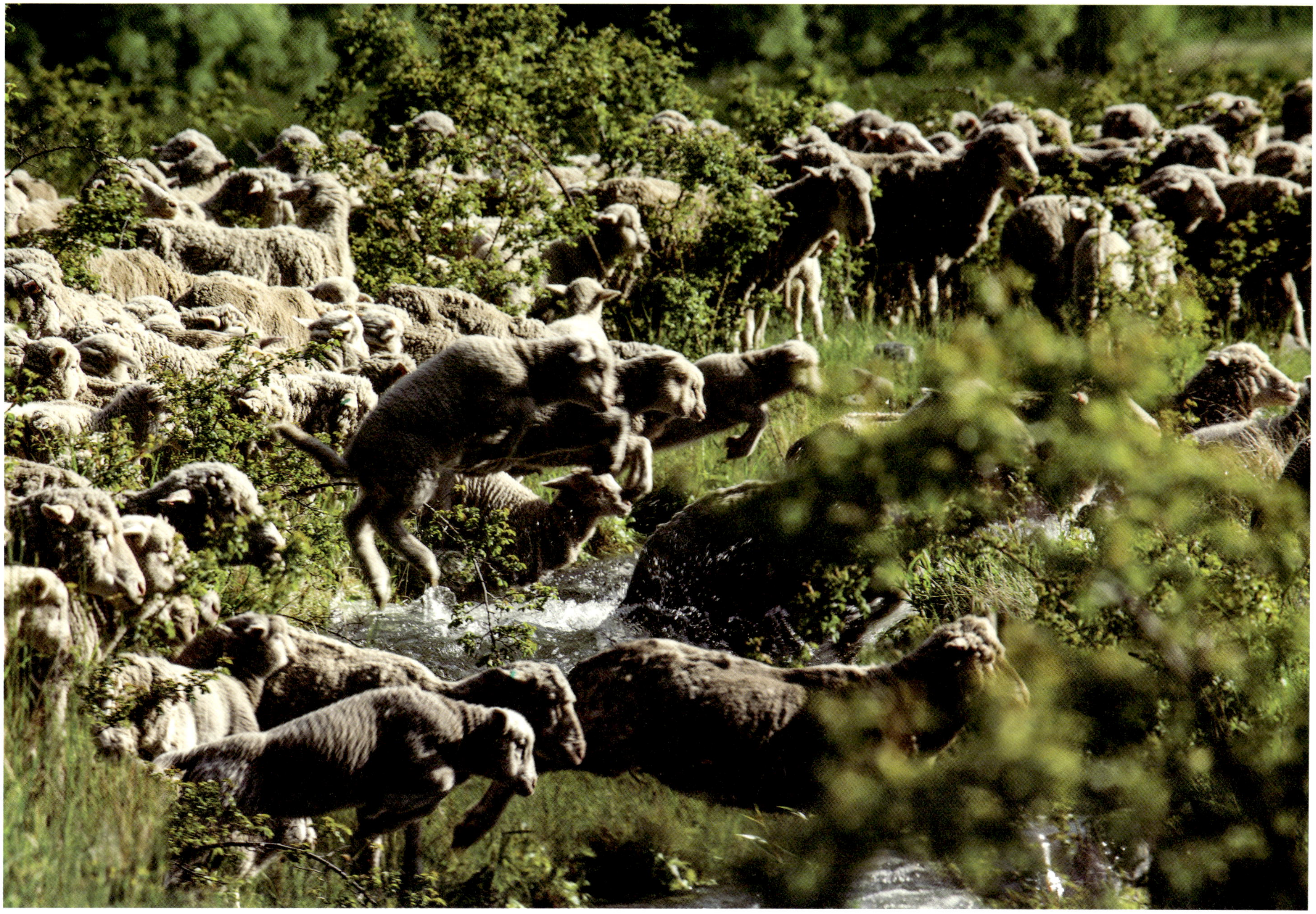

This was the real challenge for Loro Piana: finding a way of processing this extrafine wool in order to create a market that would support the enormous efforts of the growers committed to producing the world's highest quality raw material.

The process starts with a single bale, no more than 90–100 kilos of wool, known as the World Record Bale. It is the absolute finest of all the bales produced in Australia and New Zealand, the so-called Top Line 1PP bales. Loro Piana has developed special methods for washing, combing, dyeing and spinning those 100 kilos of wool to create a fabric worthy of truly exclusive suiting.

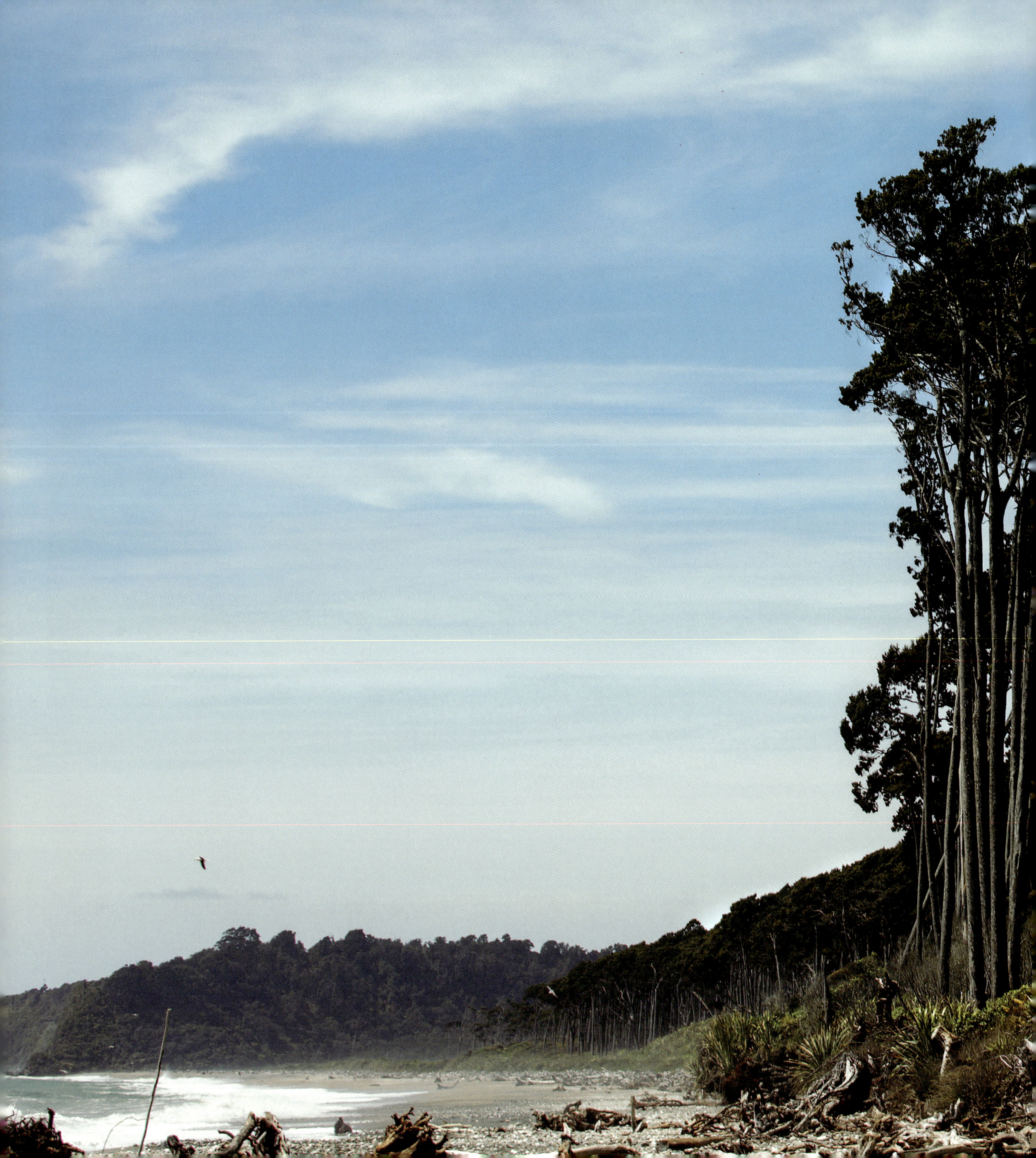

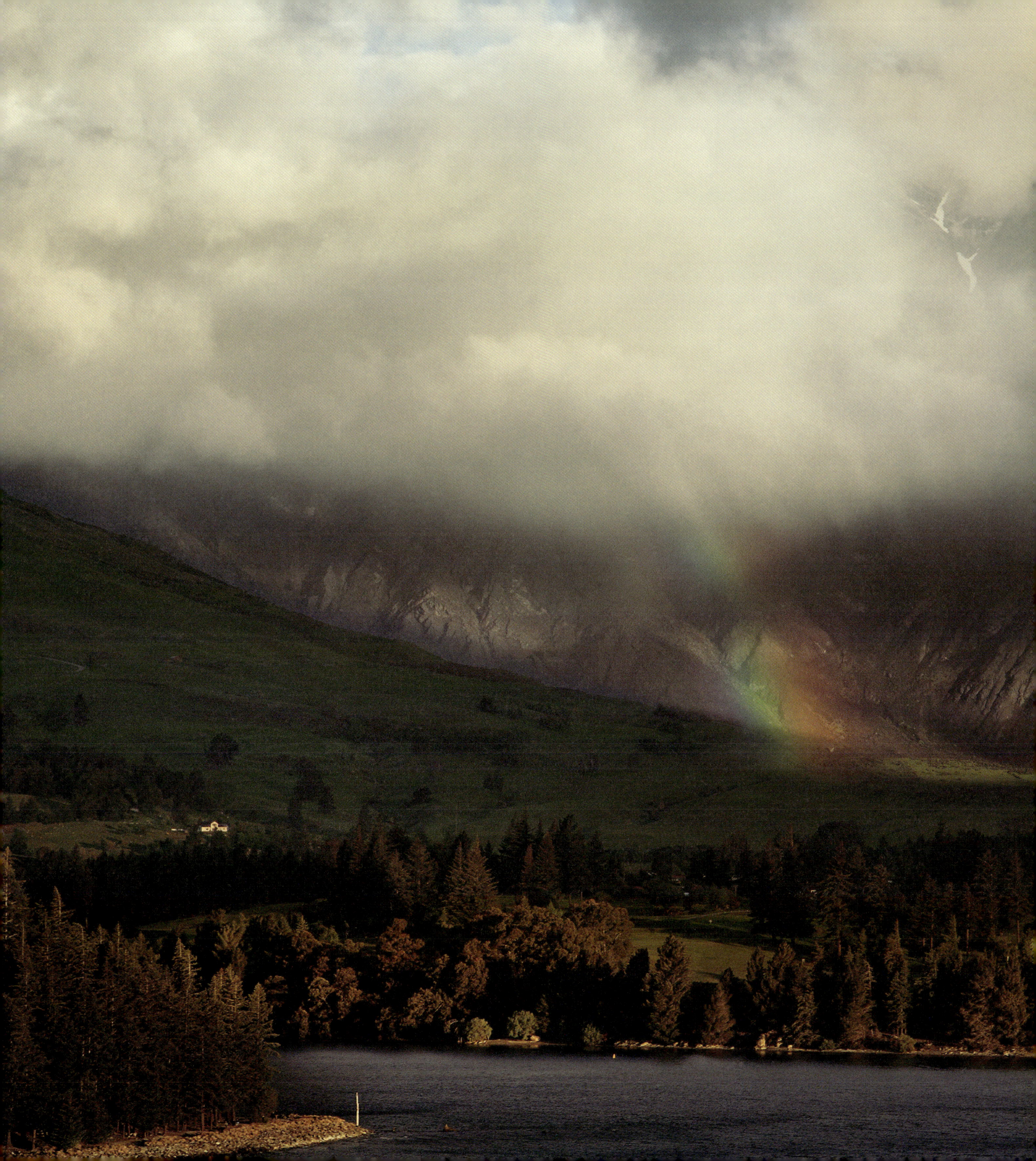

EACH SHEEP PRODUCES ABOUT

ONE KILO OF WOOL A YEAR.

THE ANIMALS ARE SHEARED IN THE

SPRING, WHEN THEY NO LONGER

NEED PROTECTION FROM

THE COLD AND RAIN, RESPECTING

THEIR NATURAL LIFE CYCLE.

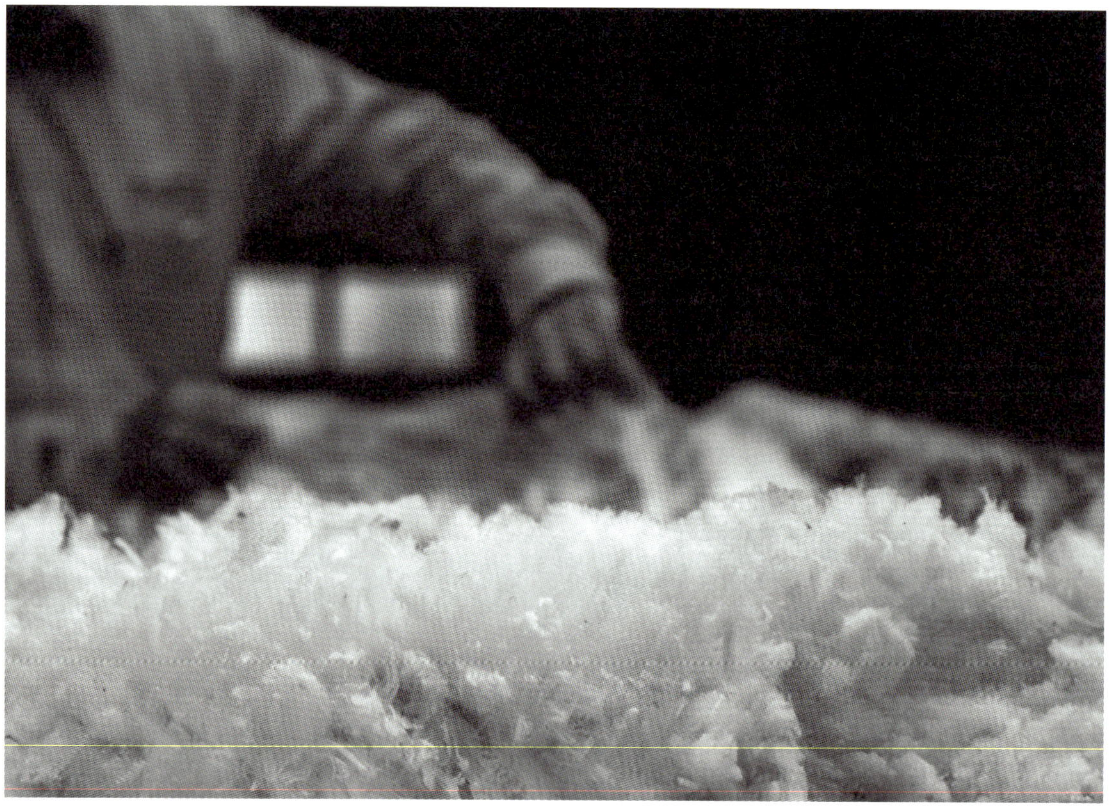

following pages
Stretching 80 kilometres, Lake Wakatipu is New Zealand's longest.

In 1997 Donald Burnett was presented with an award known as the World Wool Record Challenge Cup, now known as the Record Bale Award, which Loro Piana felt was a fitting tribute to the pioneering spirit, passion and vision of all those growers who, together with the Italian company, are constantly striving to improve the fibre. It was to be the first of many such awards that over the years have acknowledged the hard work of entire farming families in both New Zealand and Australia. These prizes are not just an end in themselves – they are a tribute to quality, passion, cooperation and the constant quest to produce wool of unthinkable fineness, down to 10 microns in diameter: the very finest animal fibre of all.

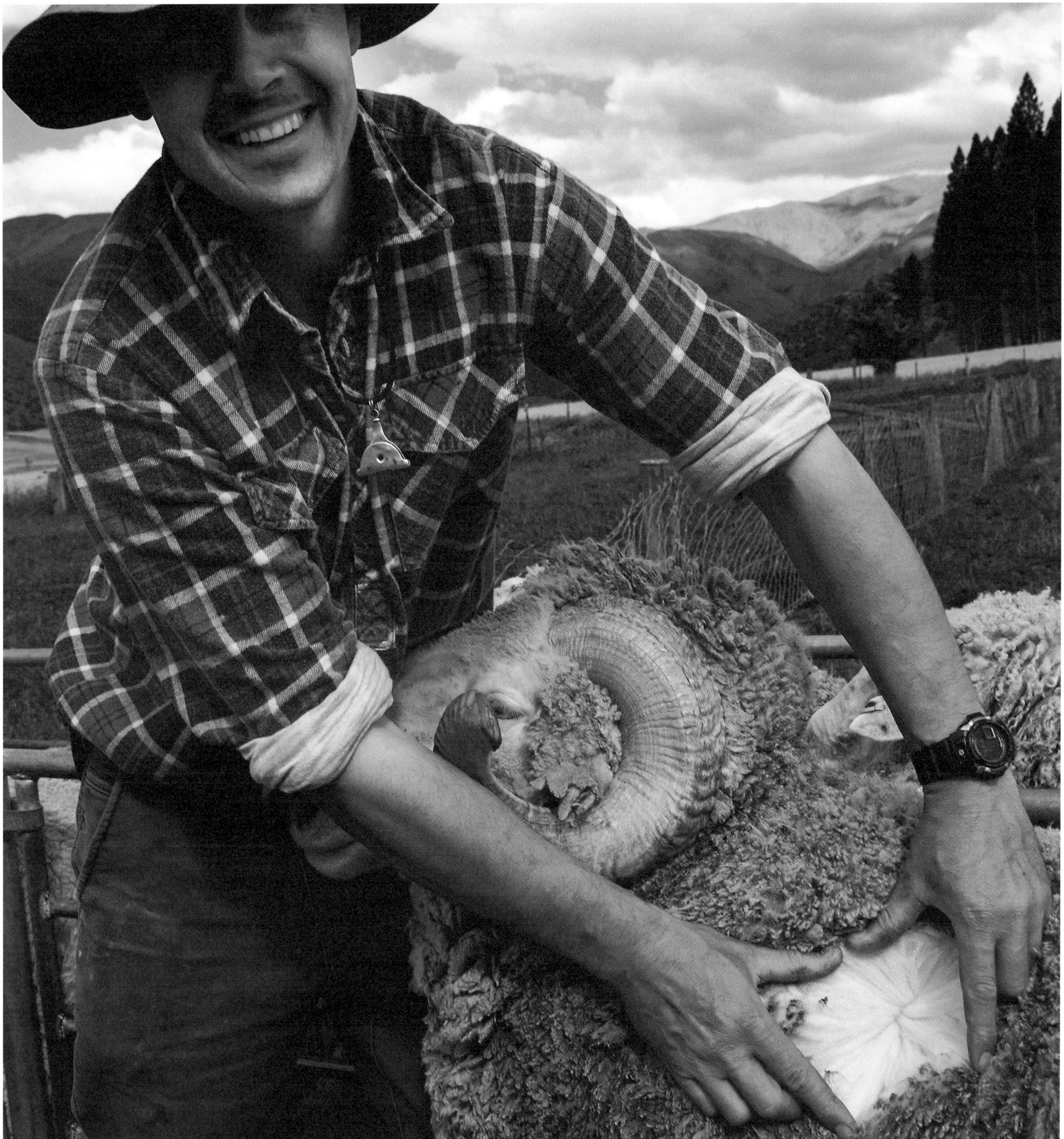

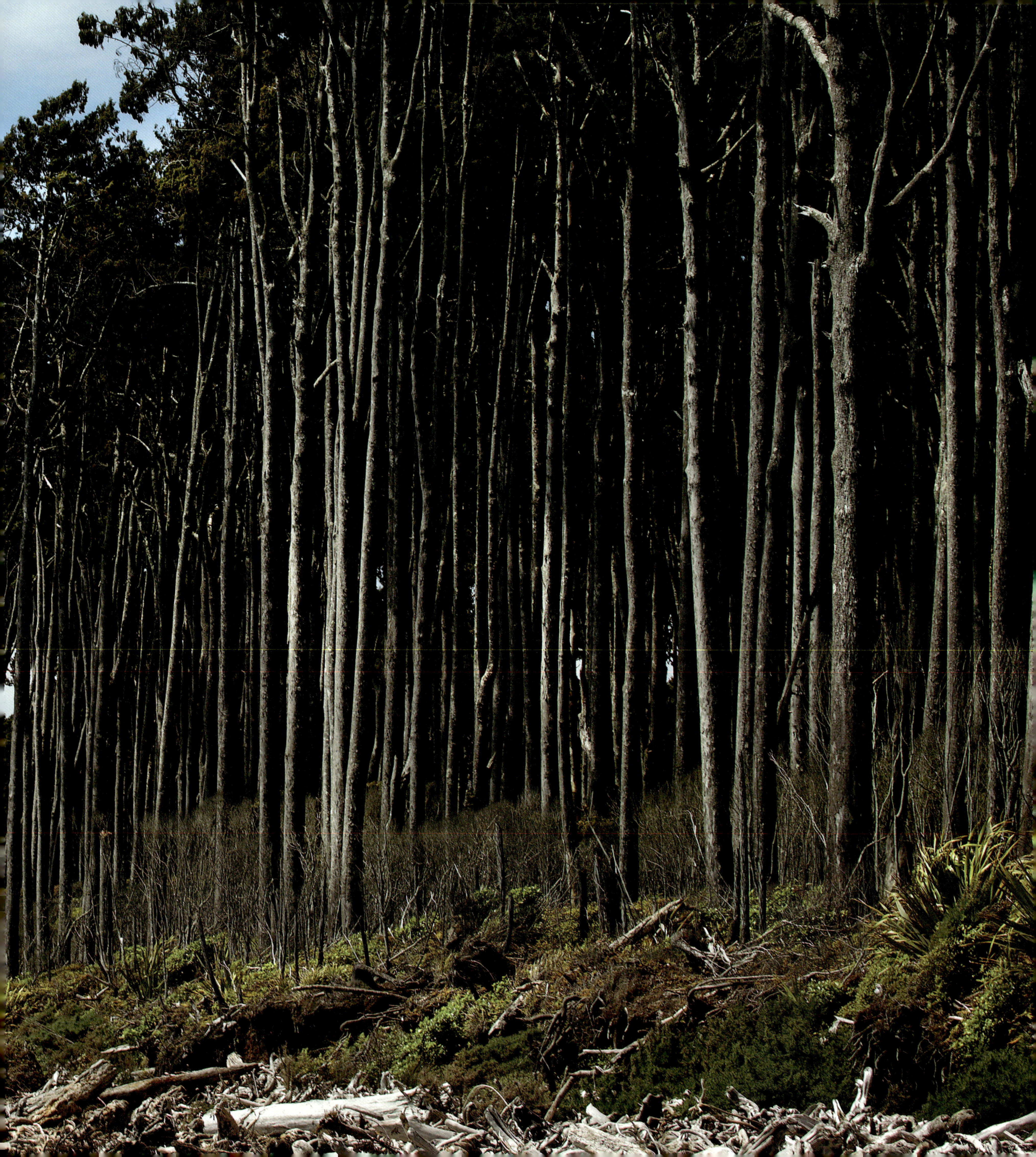

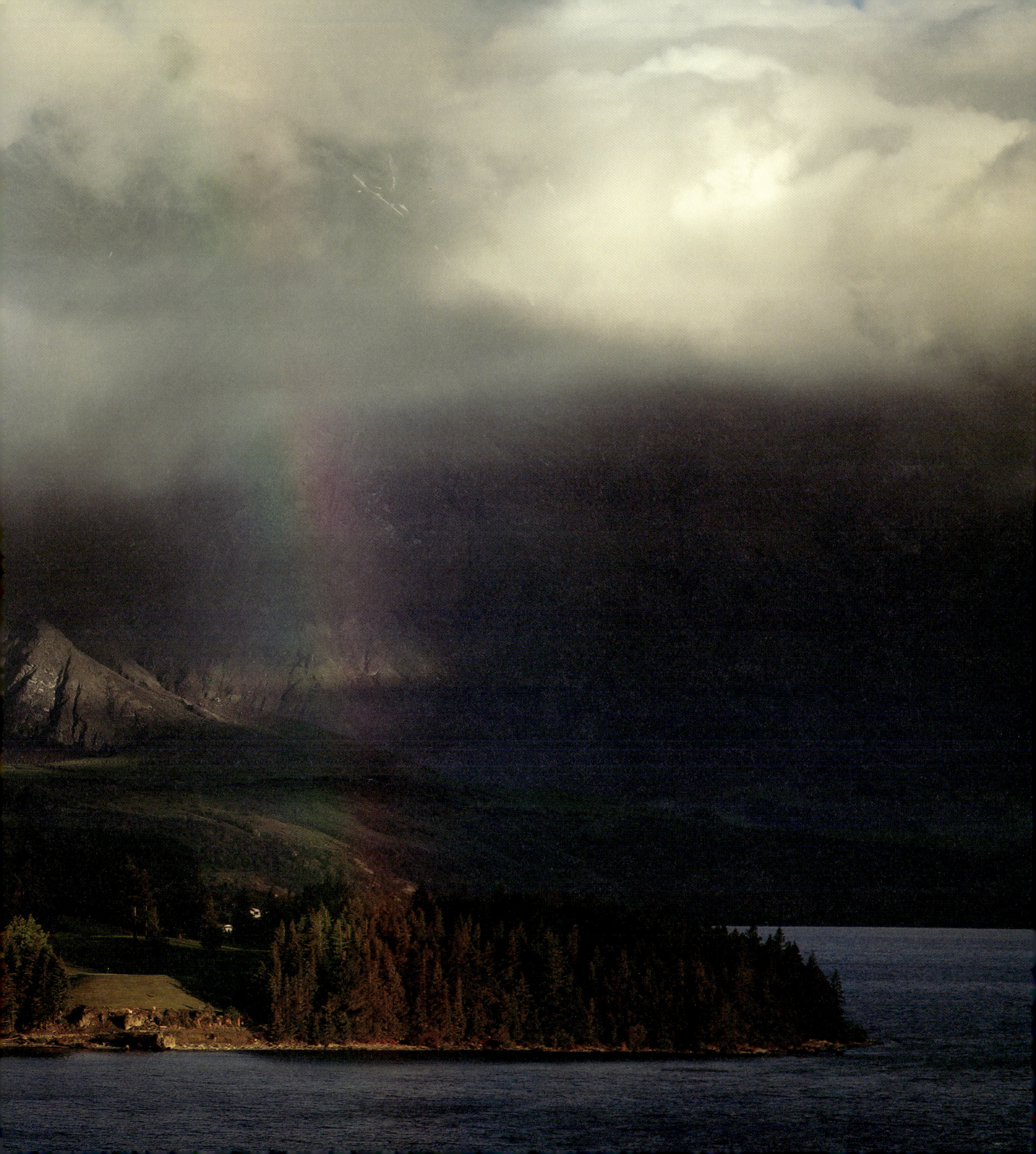

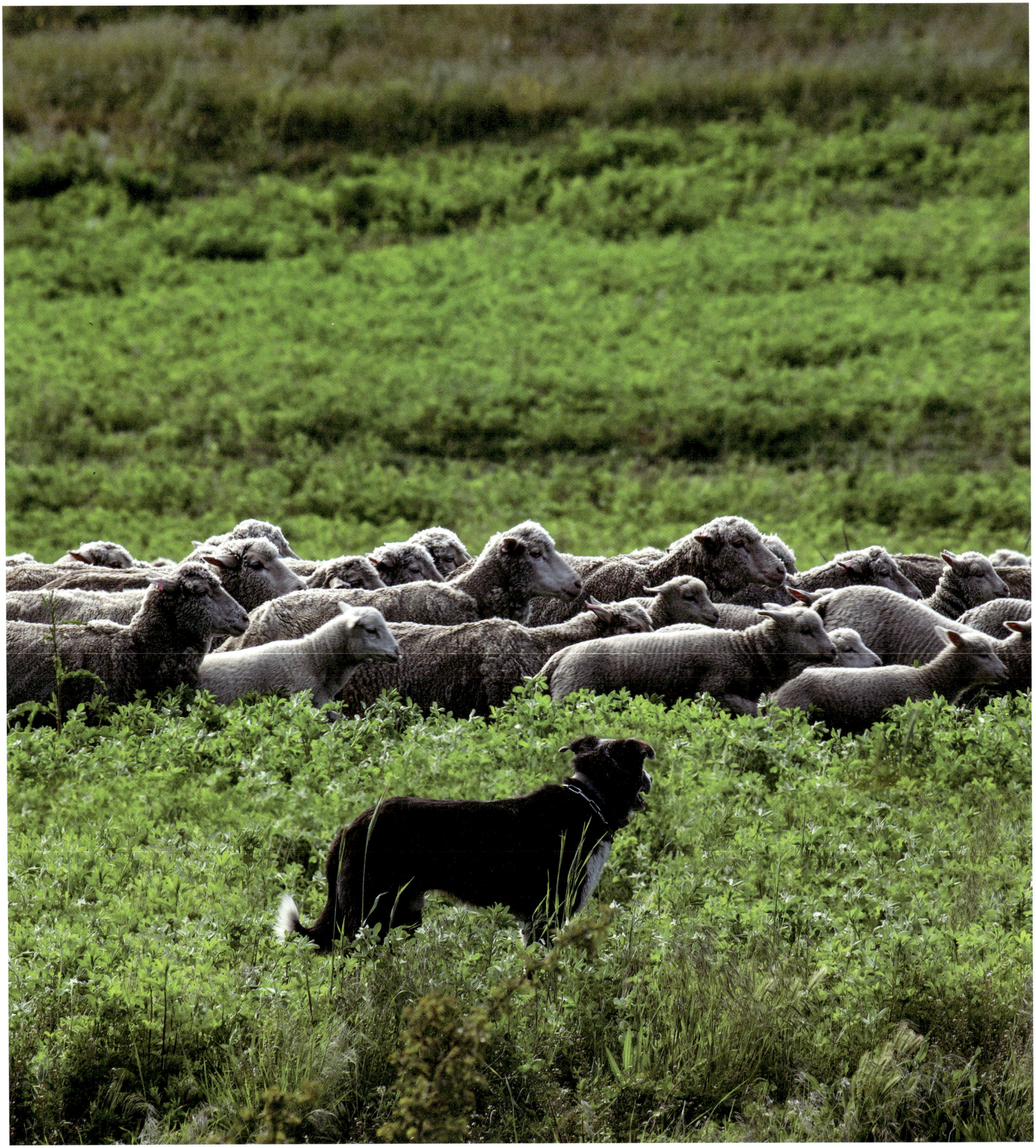

DOGS ARE WATCHFUL GUARDIANS, AN ALL IMPORTANT ROLE IN THE LIFE ON THE FARMS.

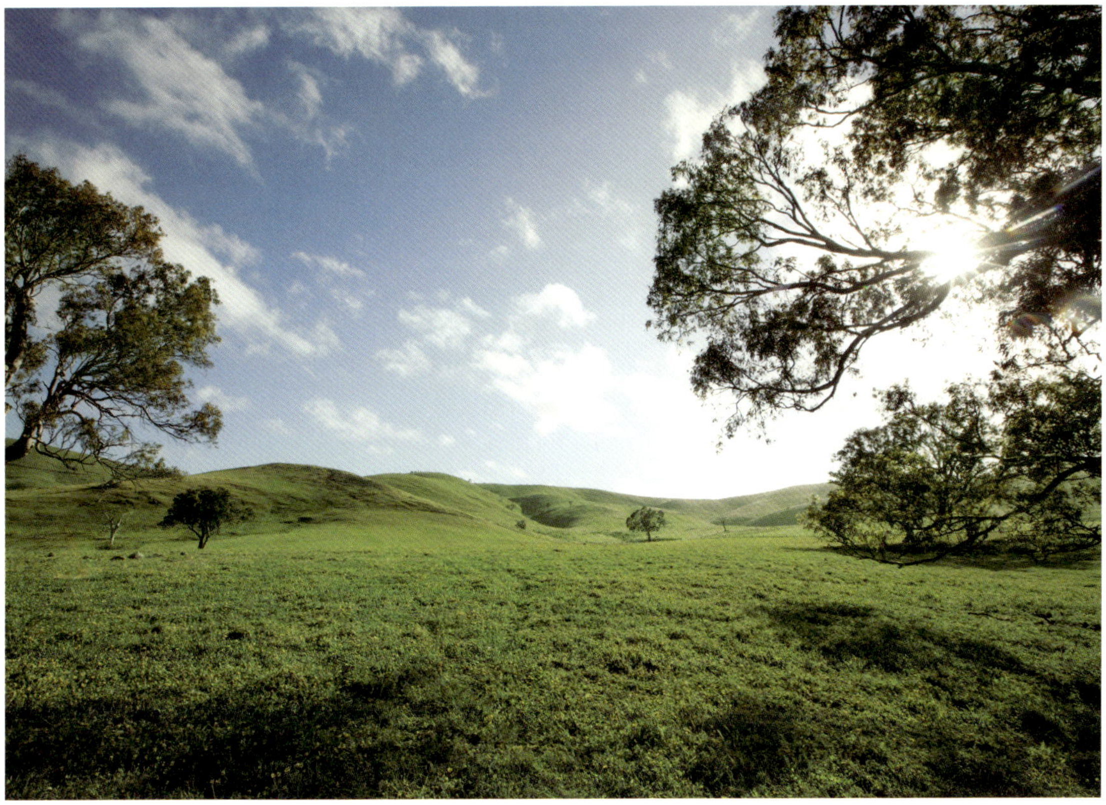

The Record Bale is also significant for another reason. During his travels across Australia and New Zealand, Pier Luigi Loro Piana realized that to produce the world's finest bales of wool, the growers were creating highly selected flocks of sheep. It was almost a new breed, this sublime merino. The wool from these flocks – from which the 100 kilos of the noblest fibre was carefully picked – was extraordinary in its own right: exquisite, superlative quality fibre measuring around 12 microns in diameter. This prompted Loro Piana's second great intuition: to buy all of the wool produced by these elite flocks and use it to make a range of fine, crease-resistant, durable, ultra-light products, warm in the winter and cool in the summer. This is how The Gift of Kings® came into being: a regal offering for those who, like Loro Piana, are passionate about true excellence.

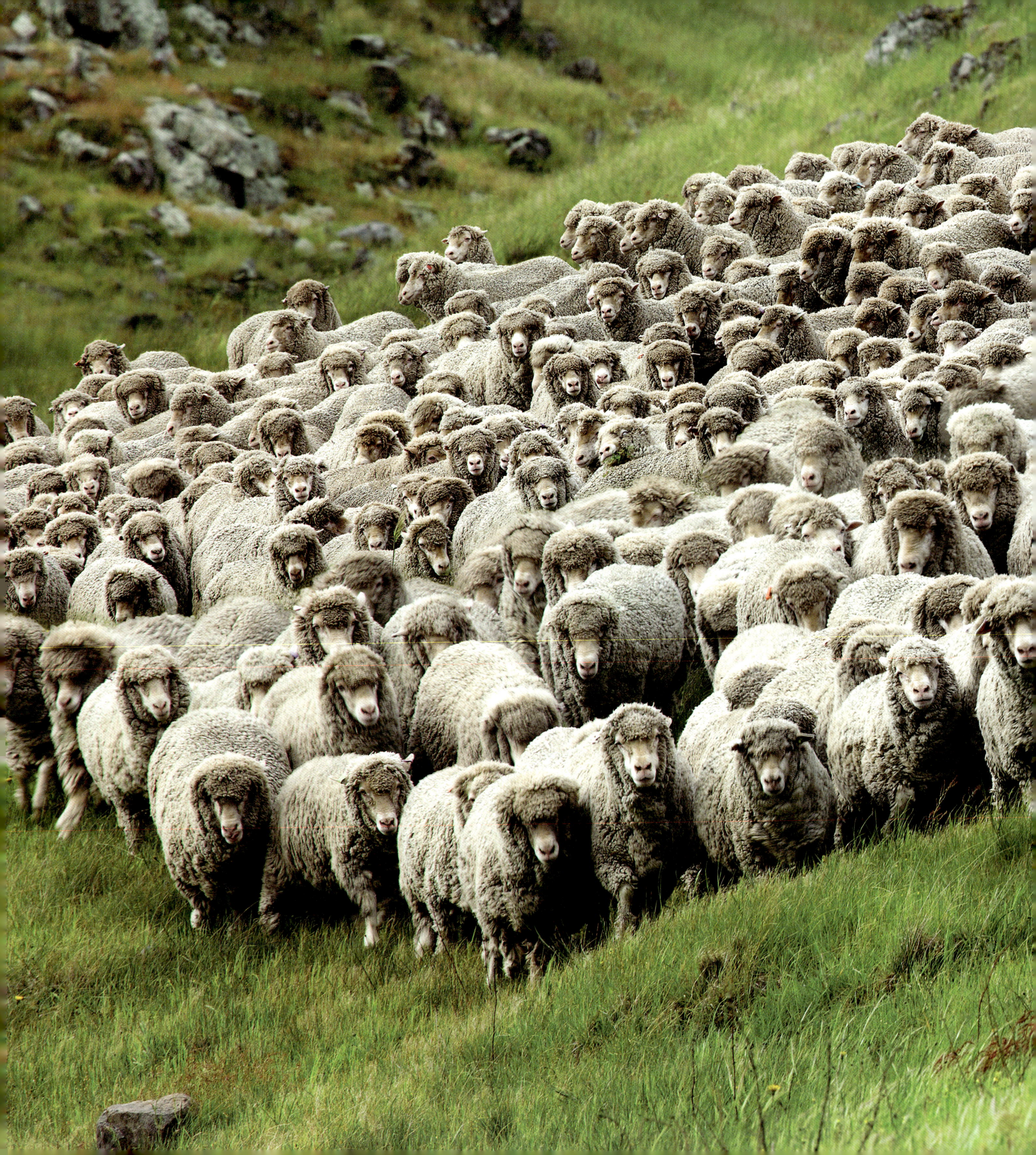

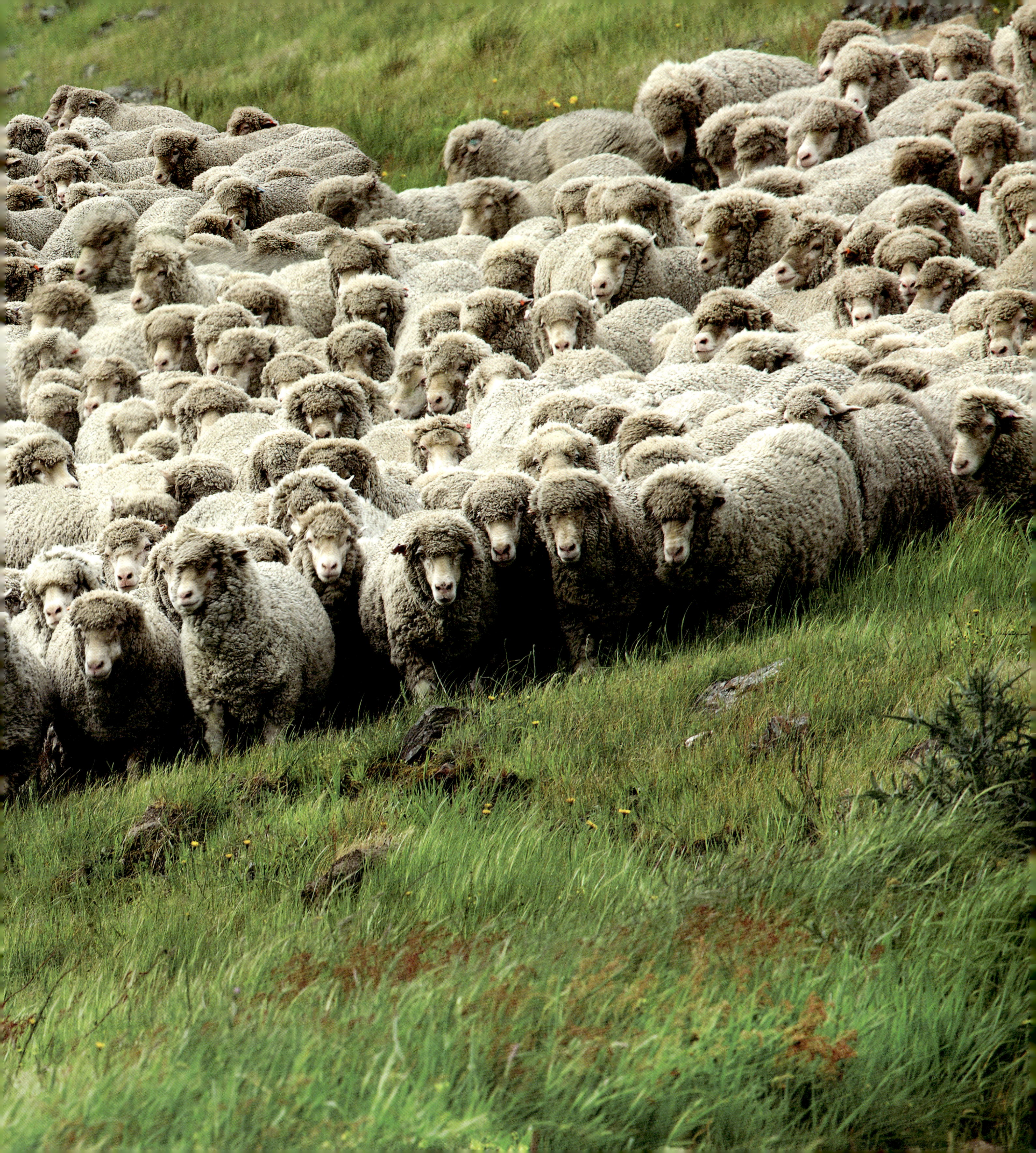

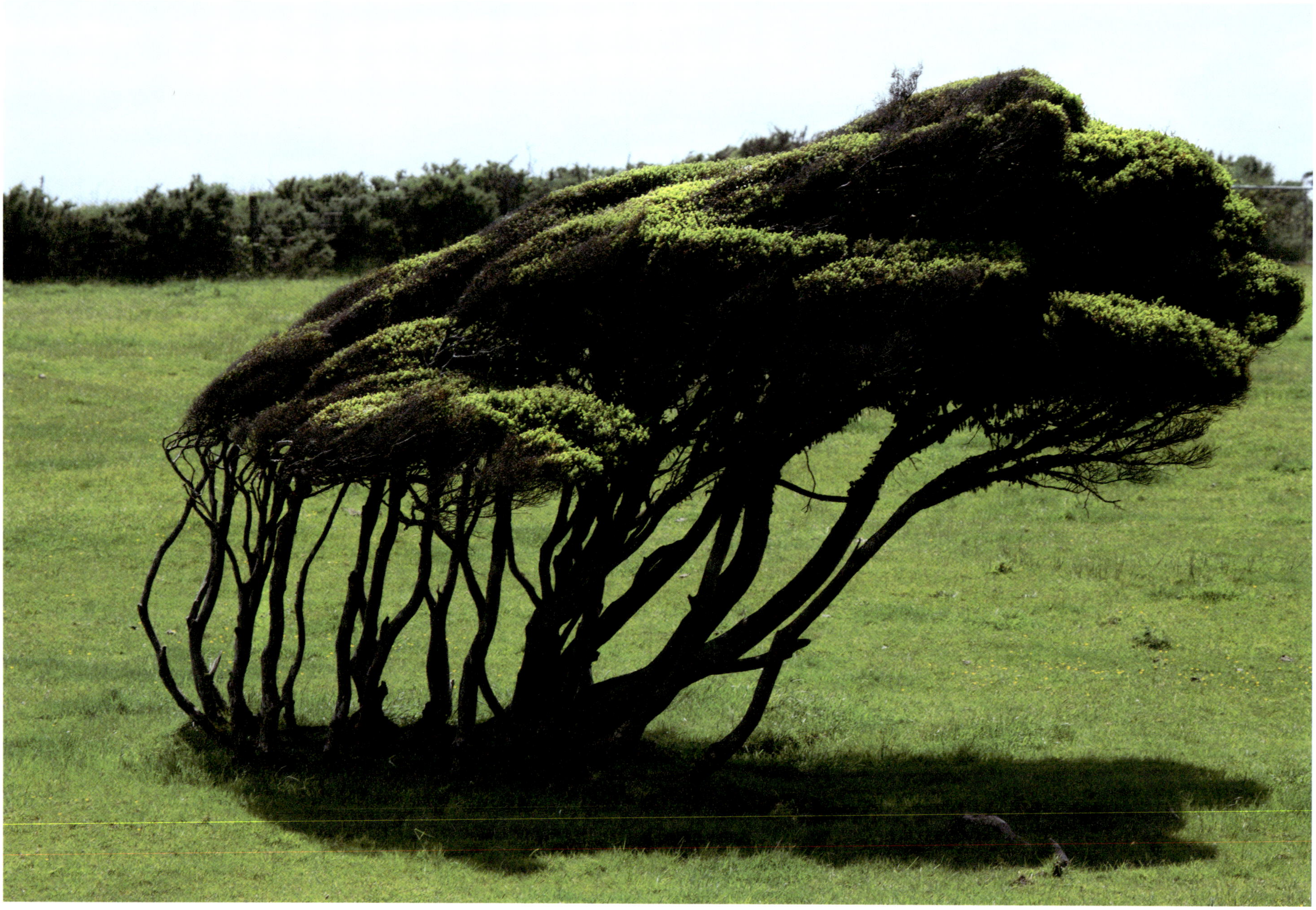

The natural landscape around
the turquoise waters of Lake Tekapo,
New Zealand.

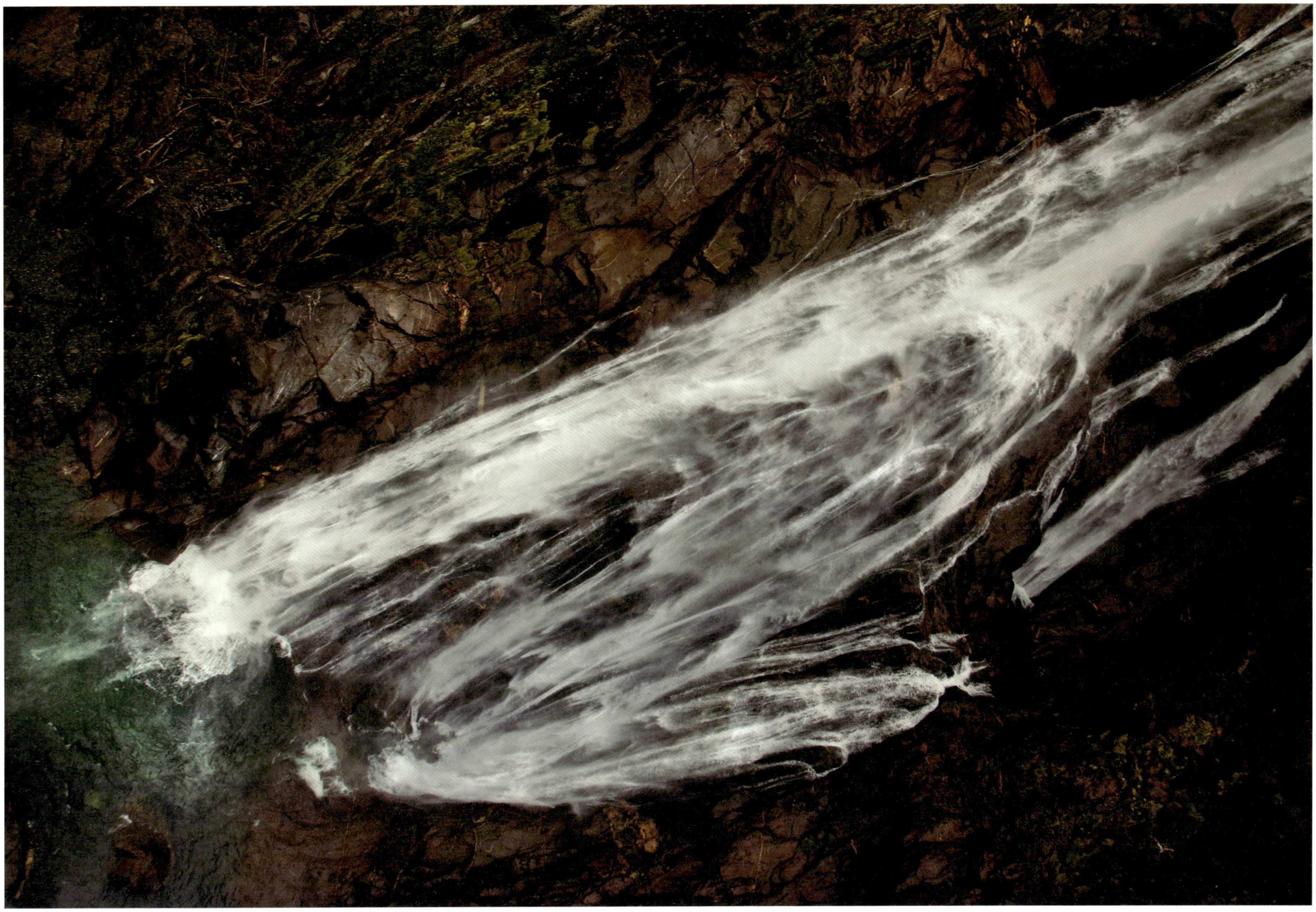

The distinctive structure of the fibre makes the garments breathable and capable of maintaining the body's natural temperature, meaning they are suited to any season and any climate. The fibre's crimped texture, along with its unparalleled fineness, ensures it is ultrasoft to wear, like a second skin: the natural choice to clothe the human body. A closer look at these downy filaments reveals a scaly surface that is inhospitable to microorganisms, lending it antibacterial properties.

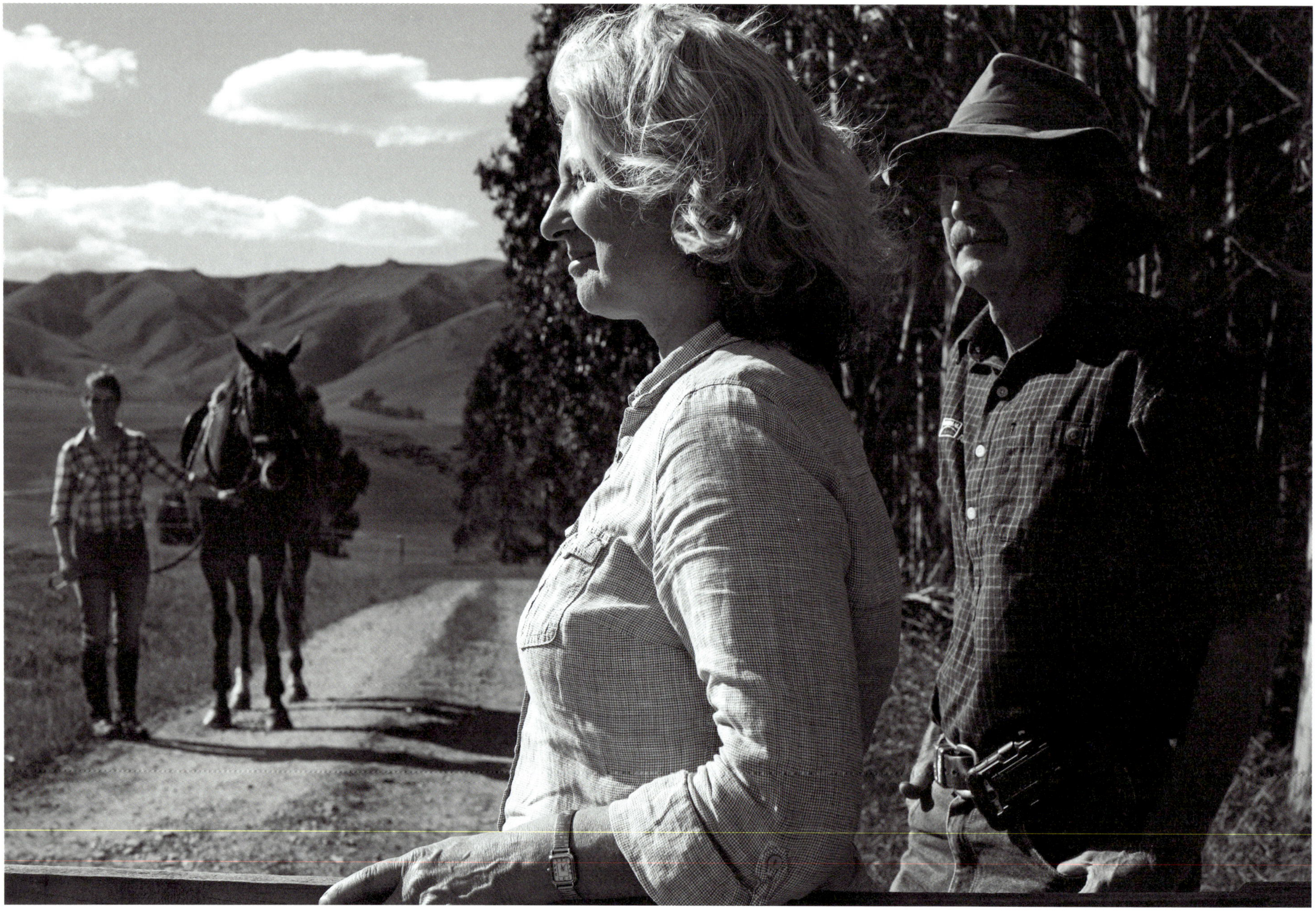

The pride of Fiona Gardner, who built
an entire flock of black sheep.
The sheep bred in ancient times
were all originally dark in colour.

following pages
The Punakaiki Pancake Rocks,
a stunning limestone rock formation
on the West Coast of New Zealand.

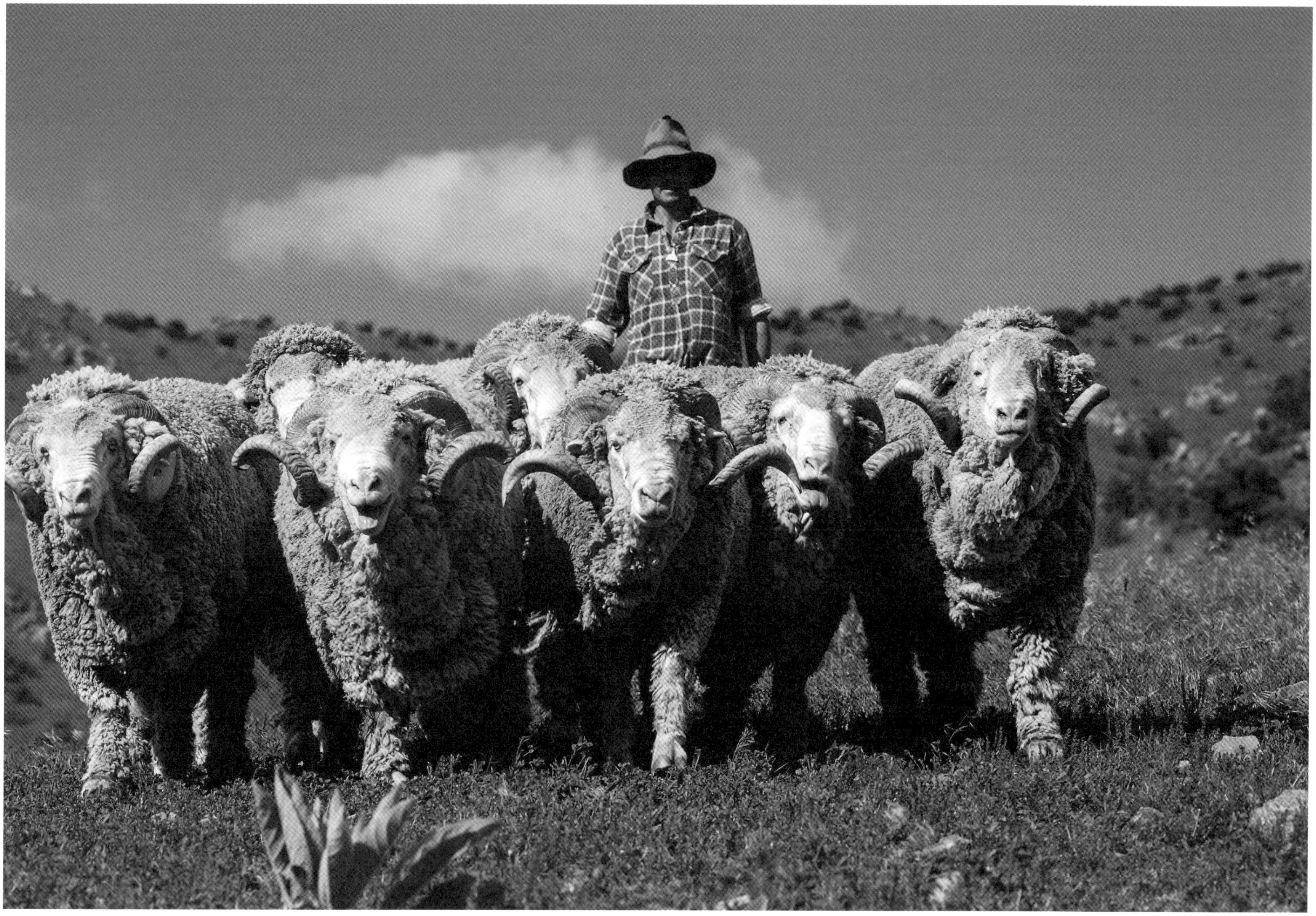

The first growers who were able to touch the Loro Piana products made with their wool, fully traceable and certified, were bowled over. And this is perhaps the most important recognition of all: the knowledge that their work is respected and honoured in unique, exceptional garments which celebrate all the properties of the fibre that they are tenaciously safeguarding and constantly improving; that these sublime expressions of their life's work can now be experienced and appreciated all over the world.

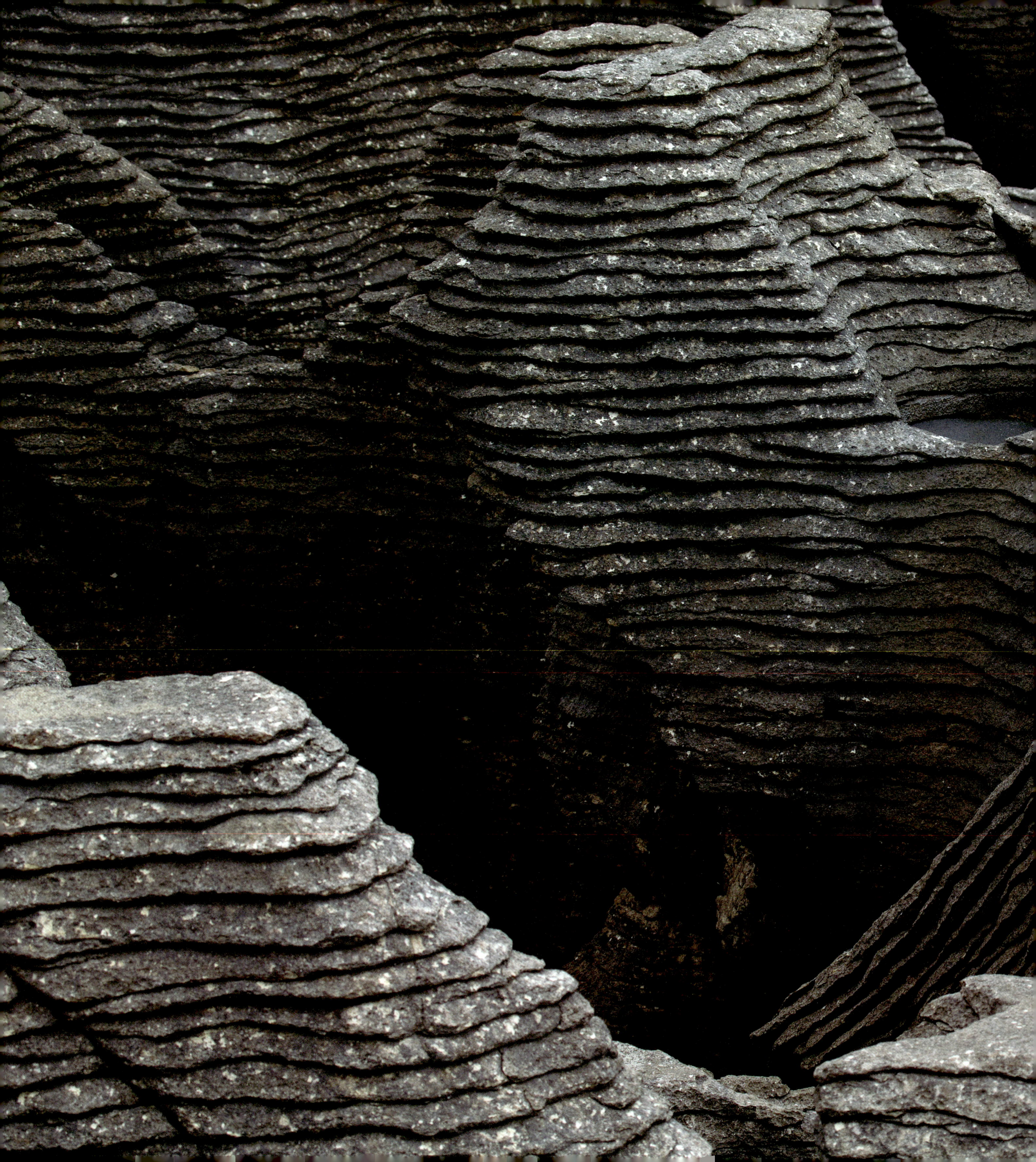

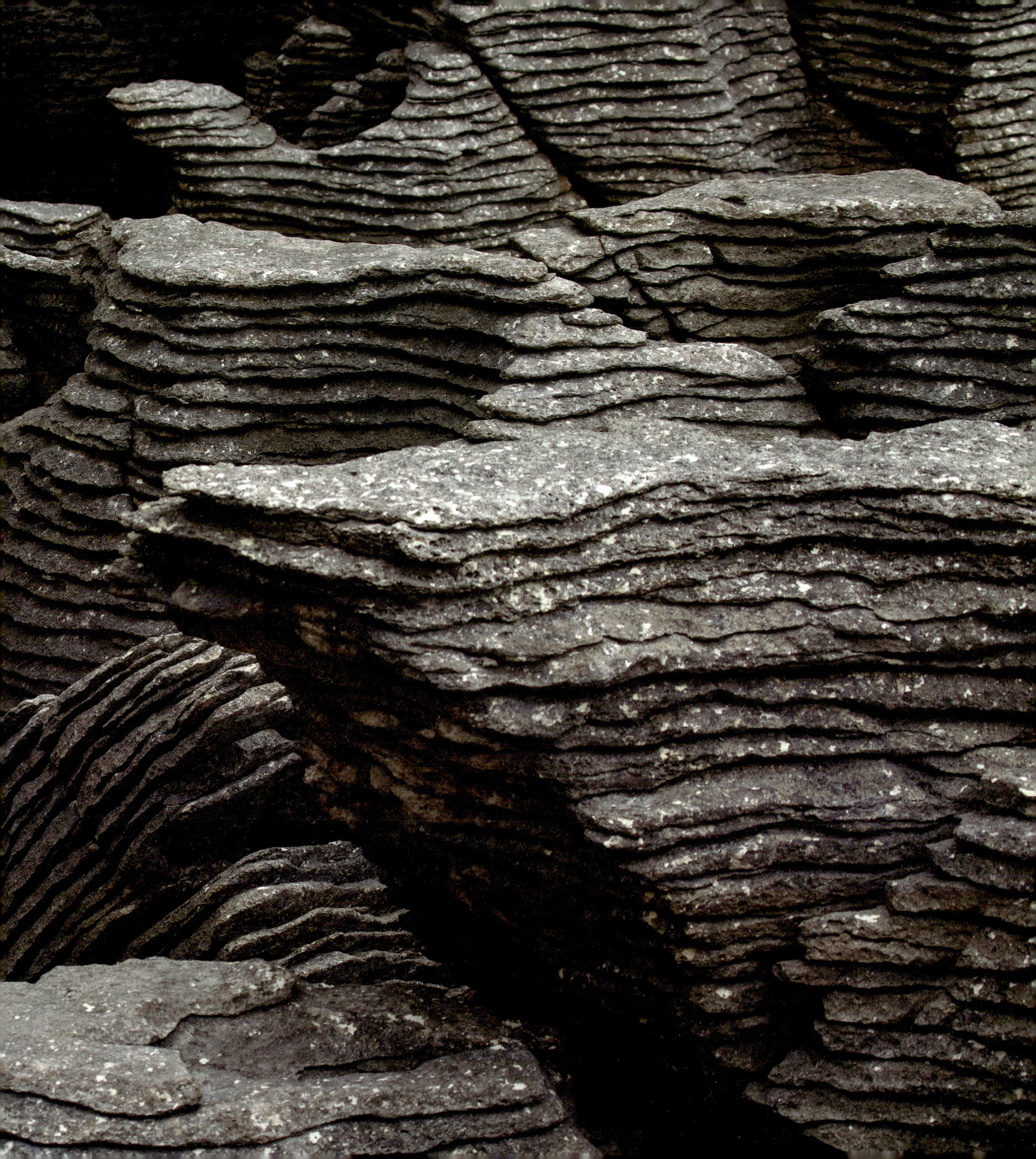

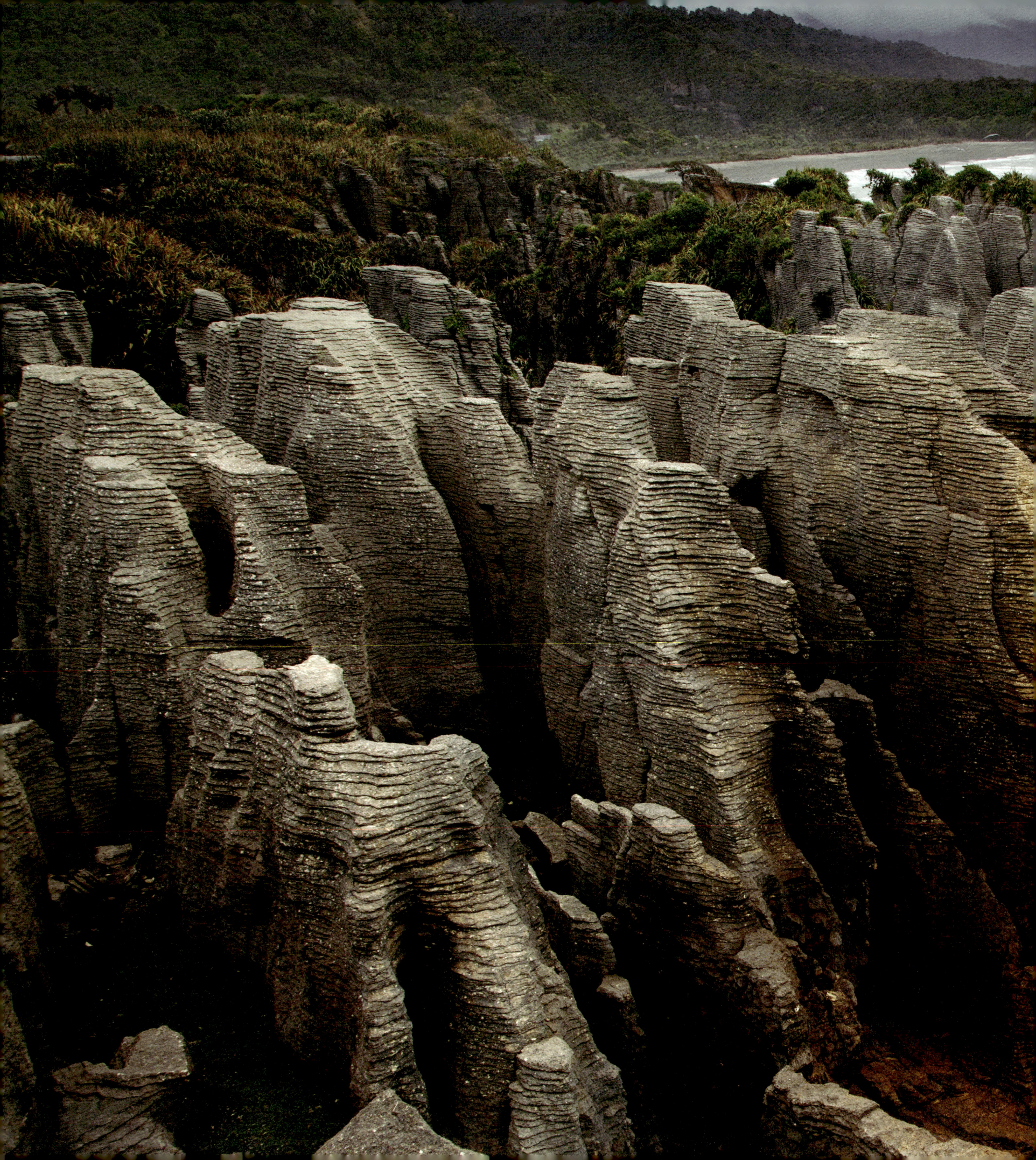

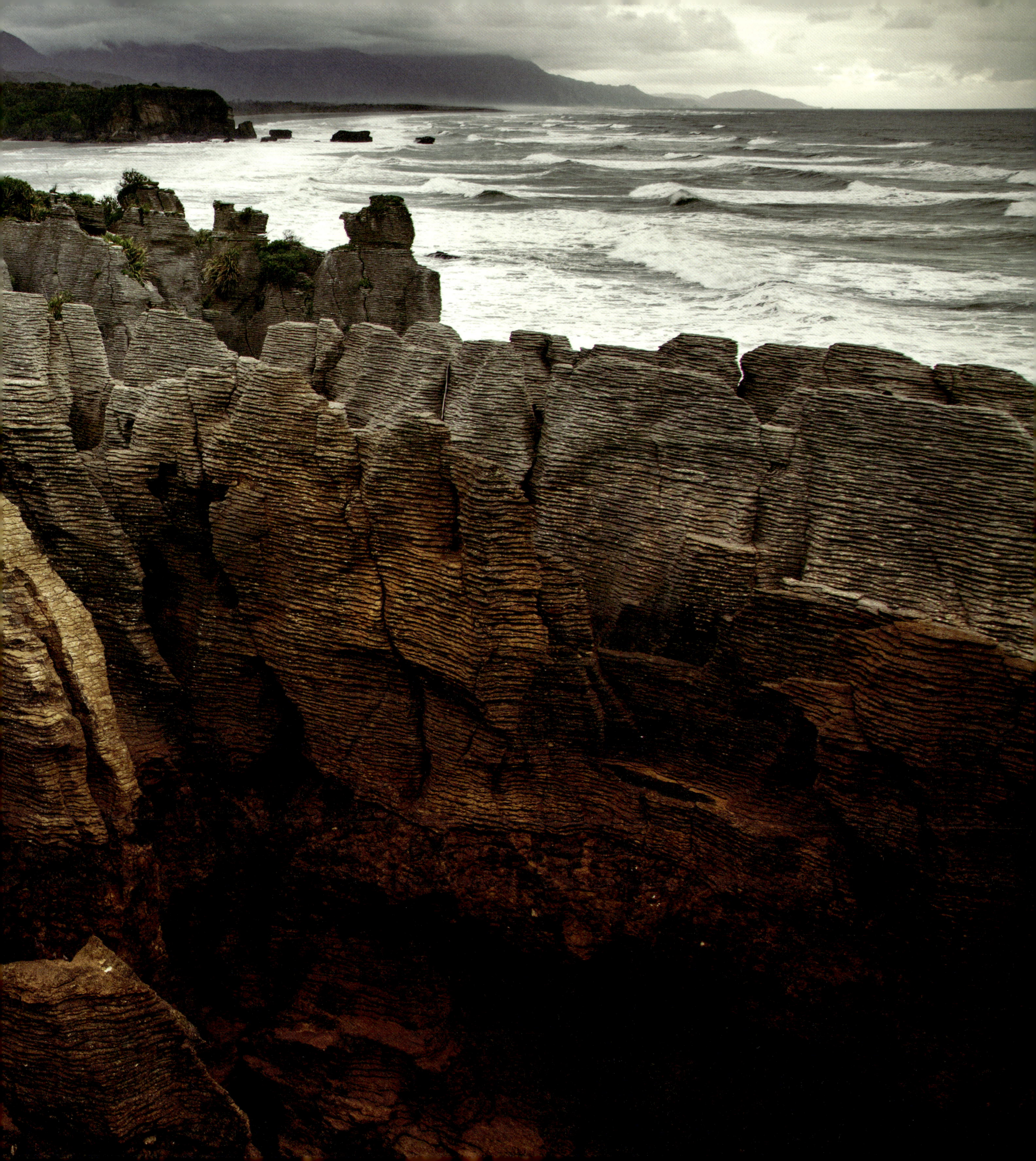

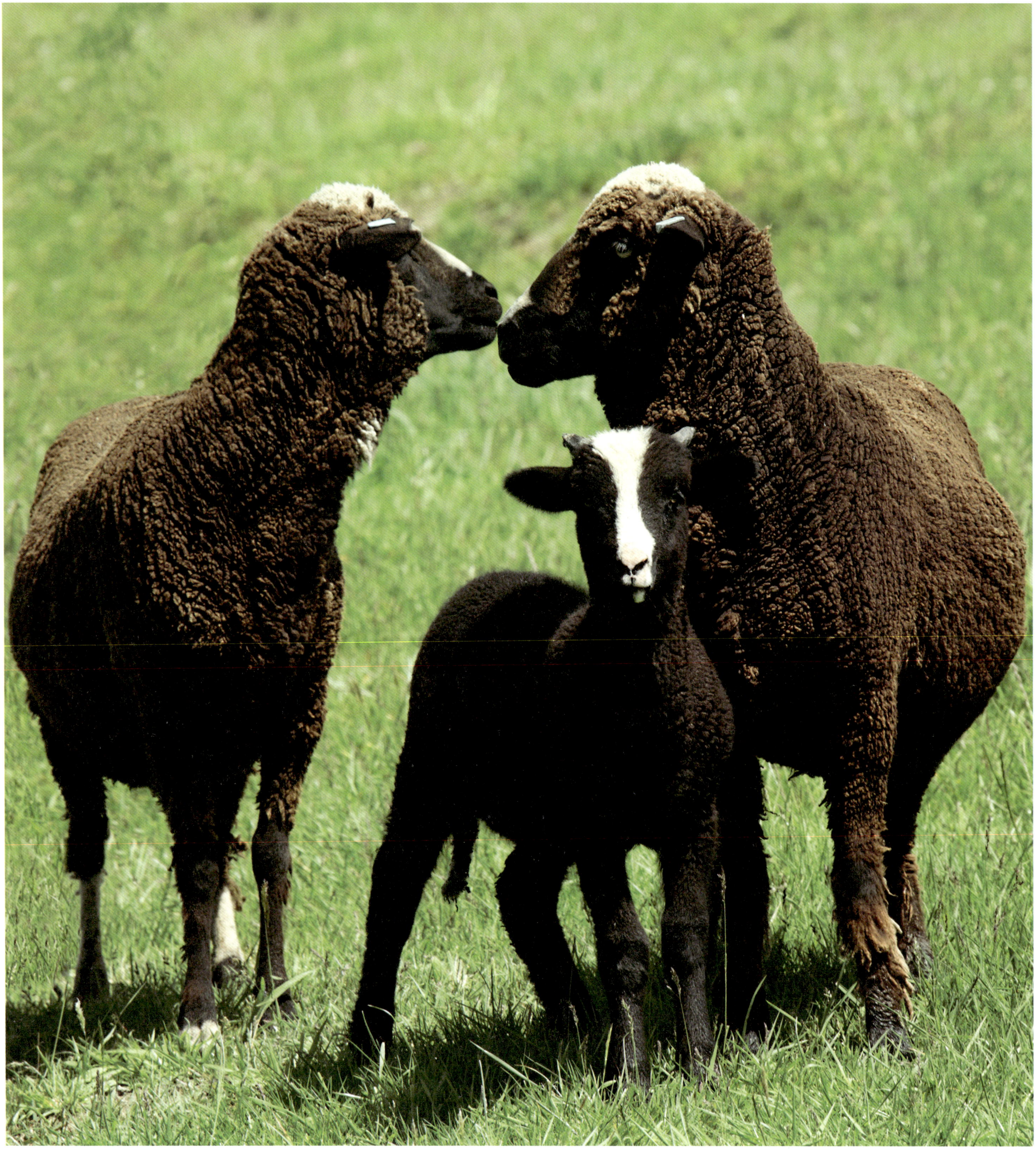

THE BEAUTIFUL BLACK MERINO SHEEP PRODUCE A FINE, "ECO-FRIENDLY" WOOL WHICH IS DARK BROWN AND NATURALLY LUSTROUS.

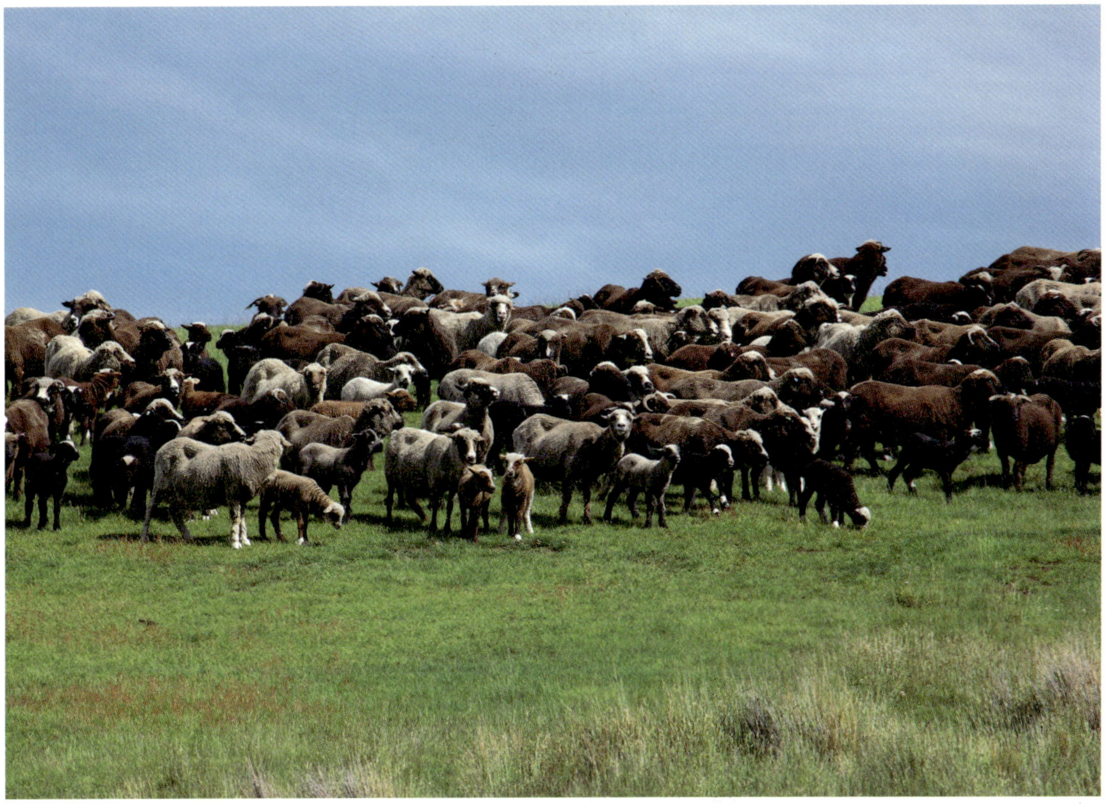

This sentiment has been captured in a photograph that Loro Piana has cherished for years, showing a proud, evidently fulfilled woman wrapped in a magnificent Loro Piana coat in a lustrous brown, outdoors in the middle of a flock of beautiful black sheep. The coat is crafted from a very special wool: the fleeces of those rare sheep. Indeed the sheep's wool was originally brown, but growers gradually selected the white variety as lighter wool is easier to dye.

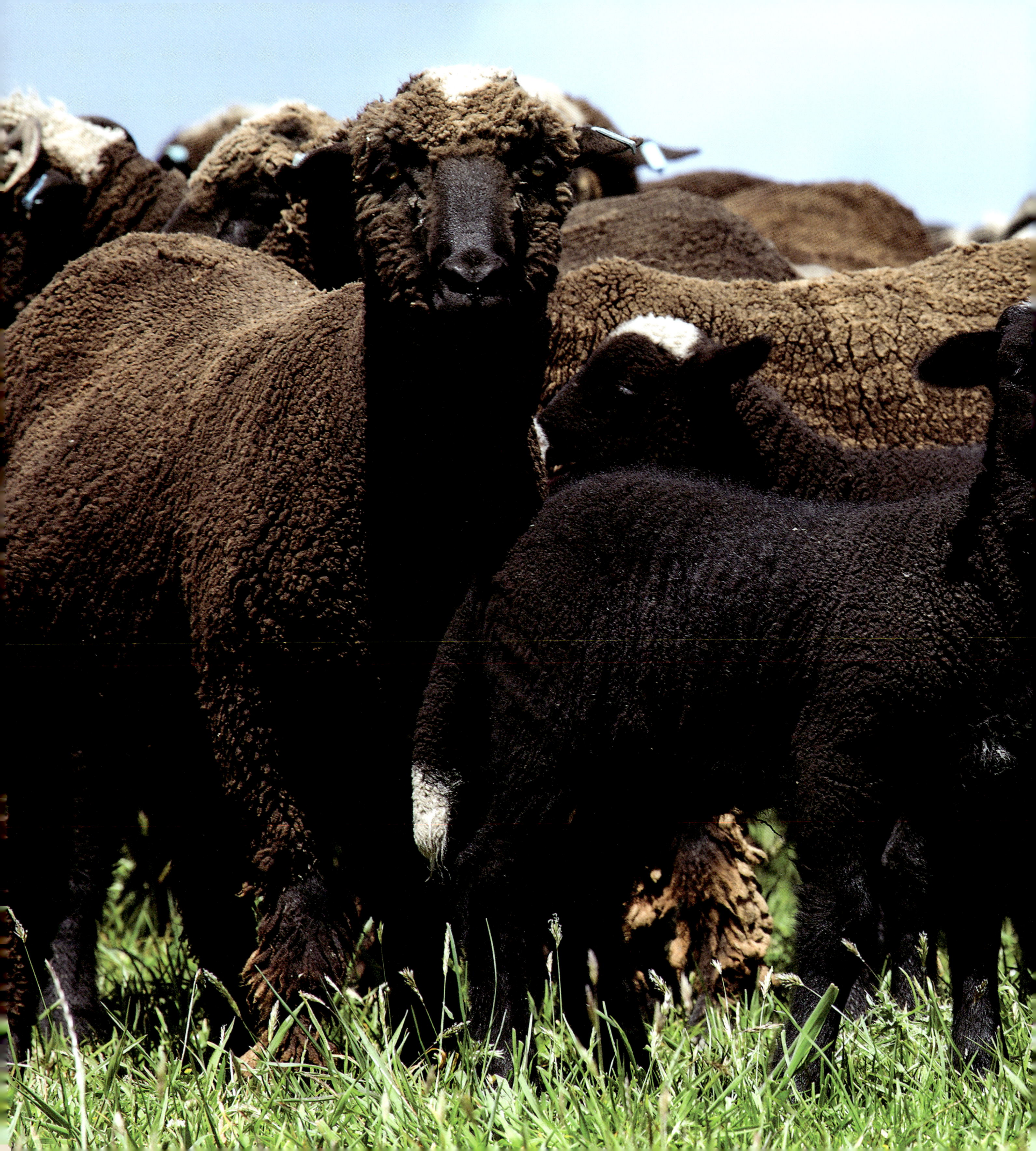

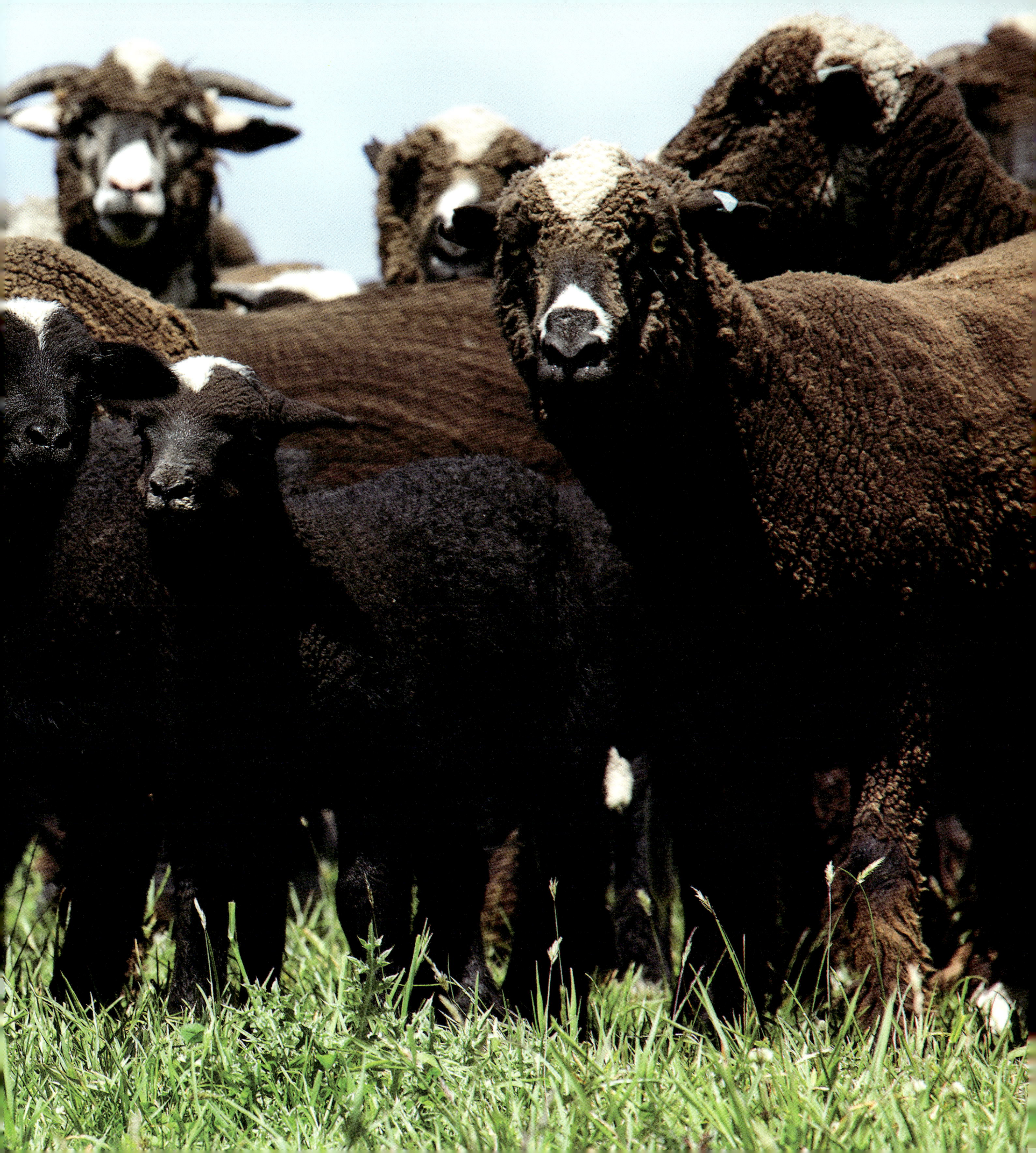

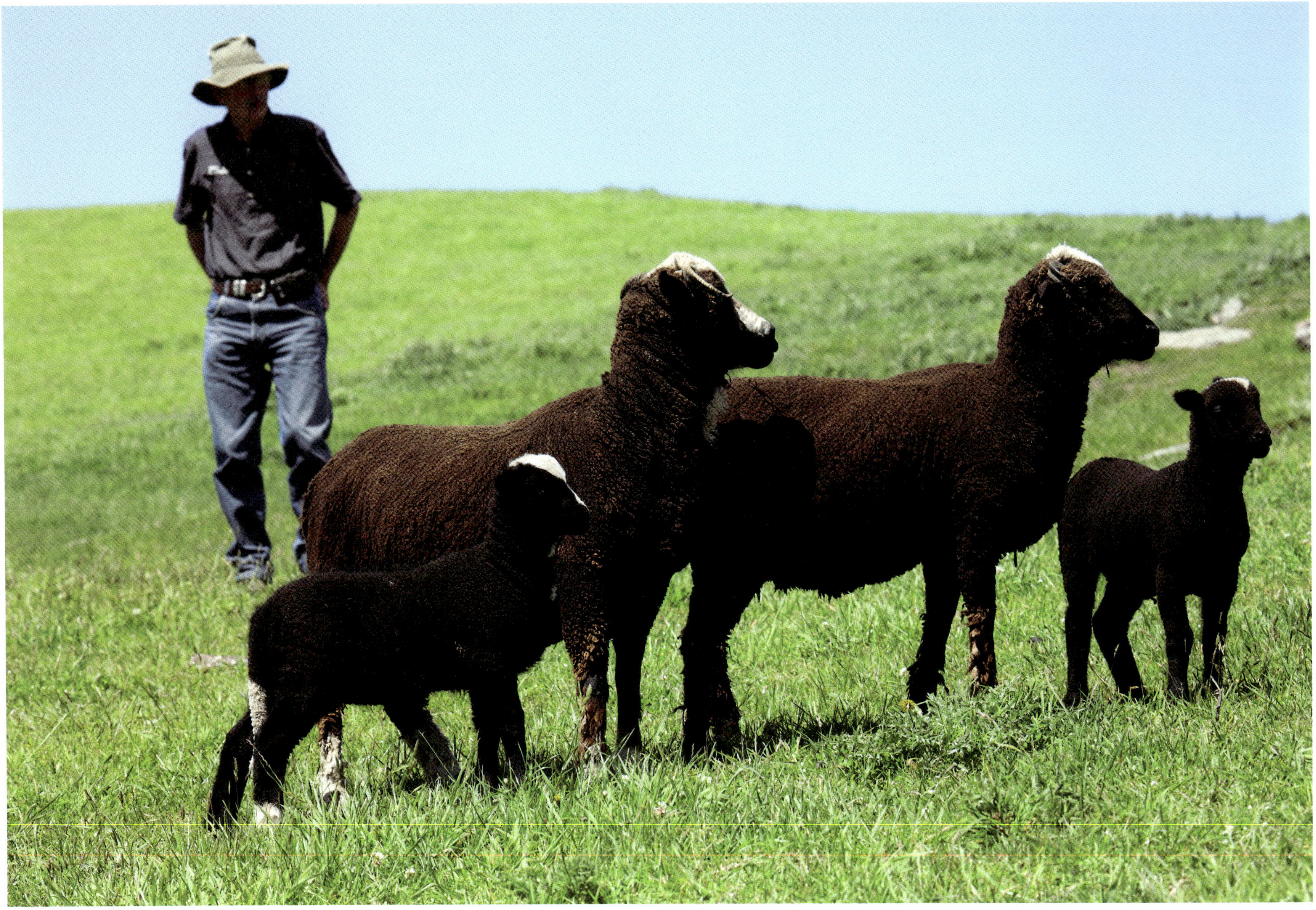

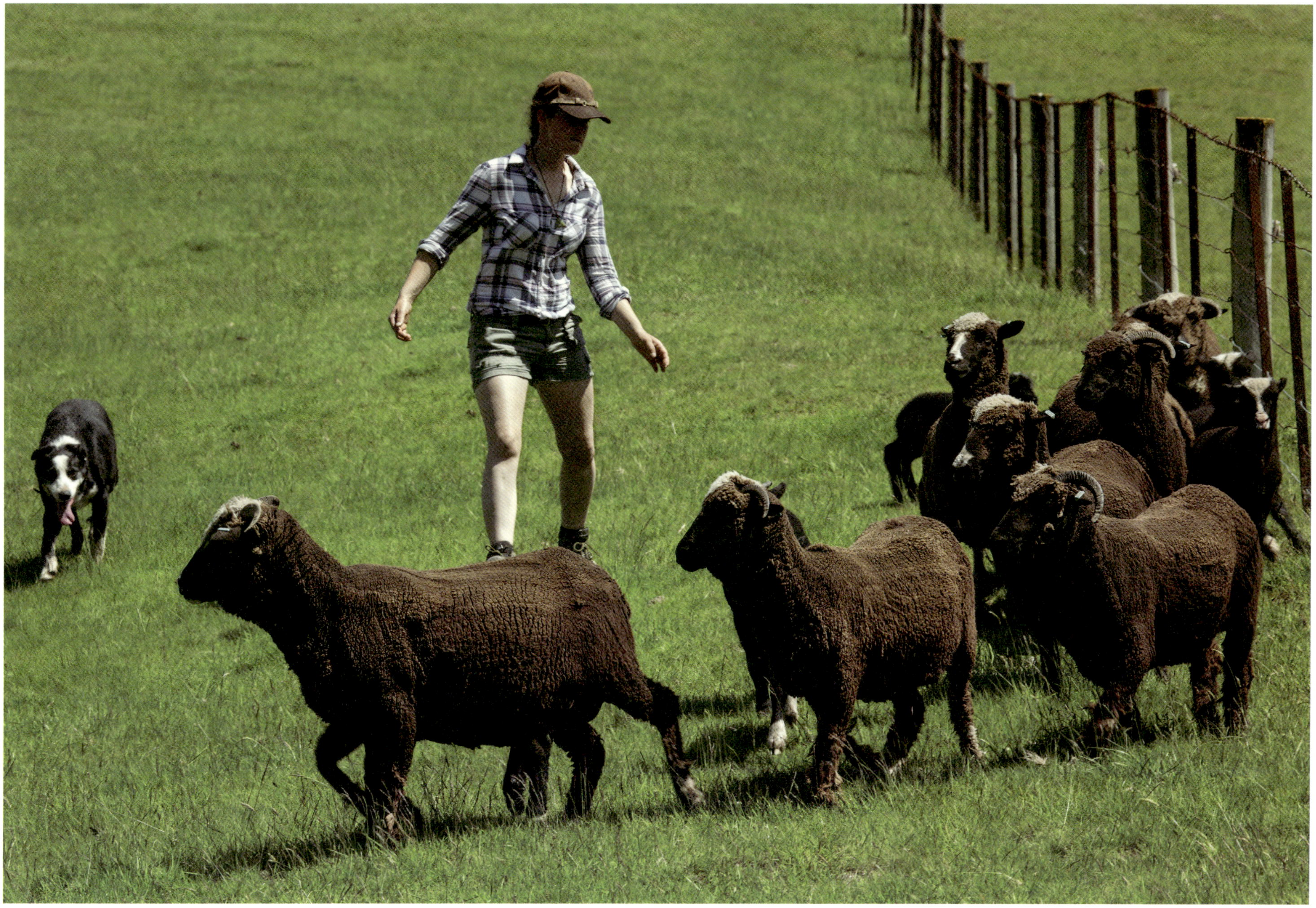

This meant that black sheep became a minority, but Fiona Gardner (the woman in the brown coat) is a dreamer, passionate about nature and its precious gifts and like Loro Piana, unable to accept the idea that an excellence of this calibre could simply die out. So she set about building a small flock of black sheep and brought this early species back from the brink. On meeting Fiona, Loro Piana realized that no dream is too bold, and vowed to back her project. The resulting product is unique, a wool with a deep, lustrous hue, not dyed; ever different and ever changing, just like nature. That photograph holds the story of a dream, a journey through time.

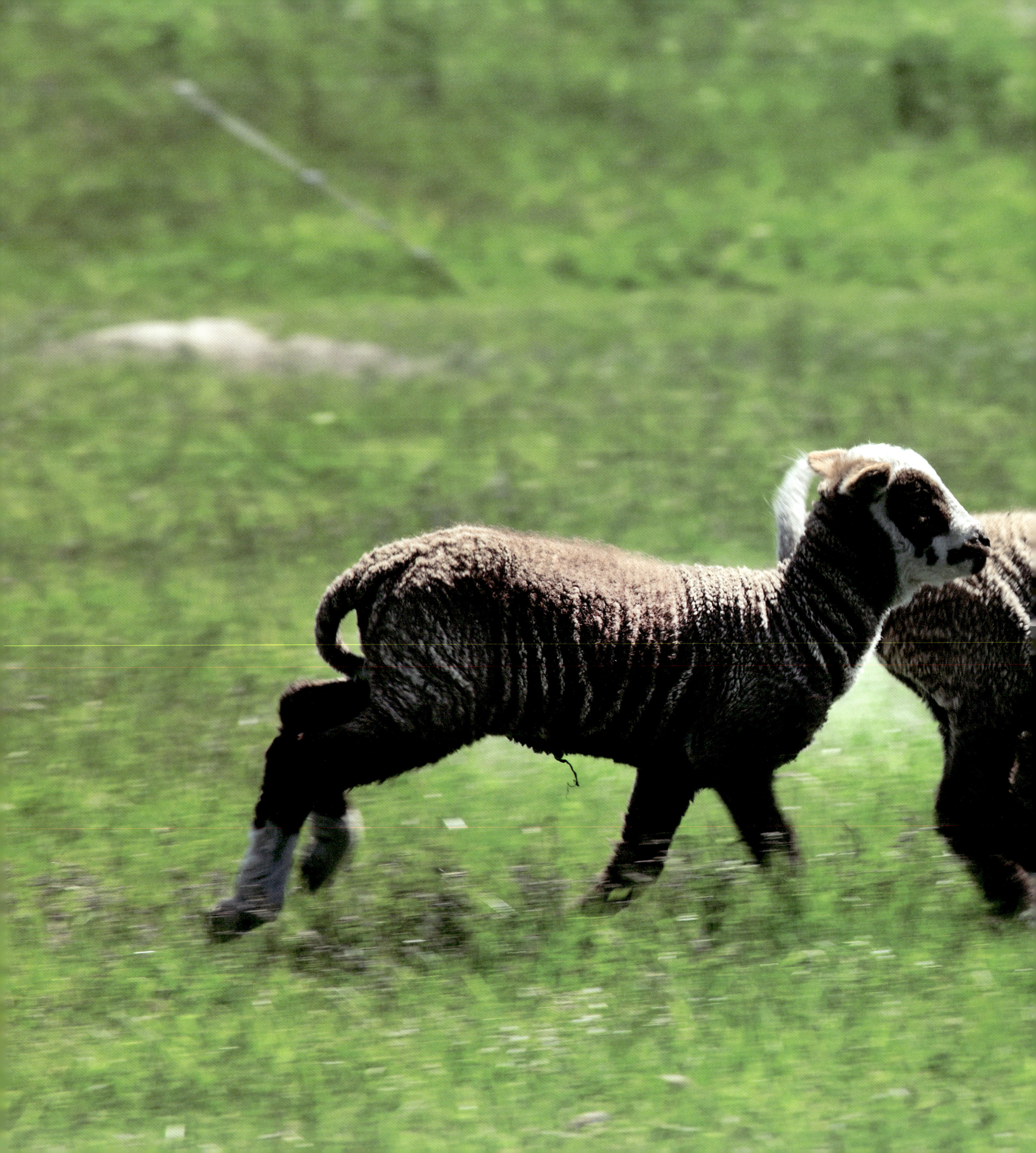

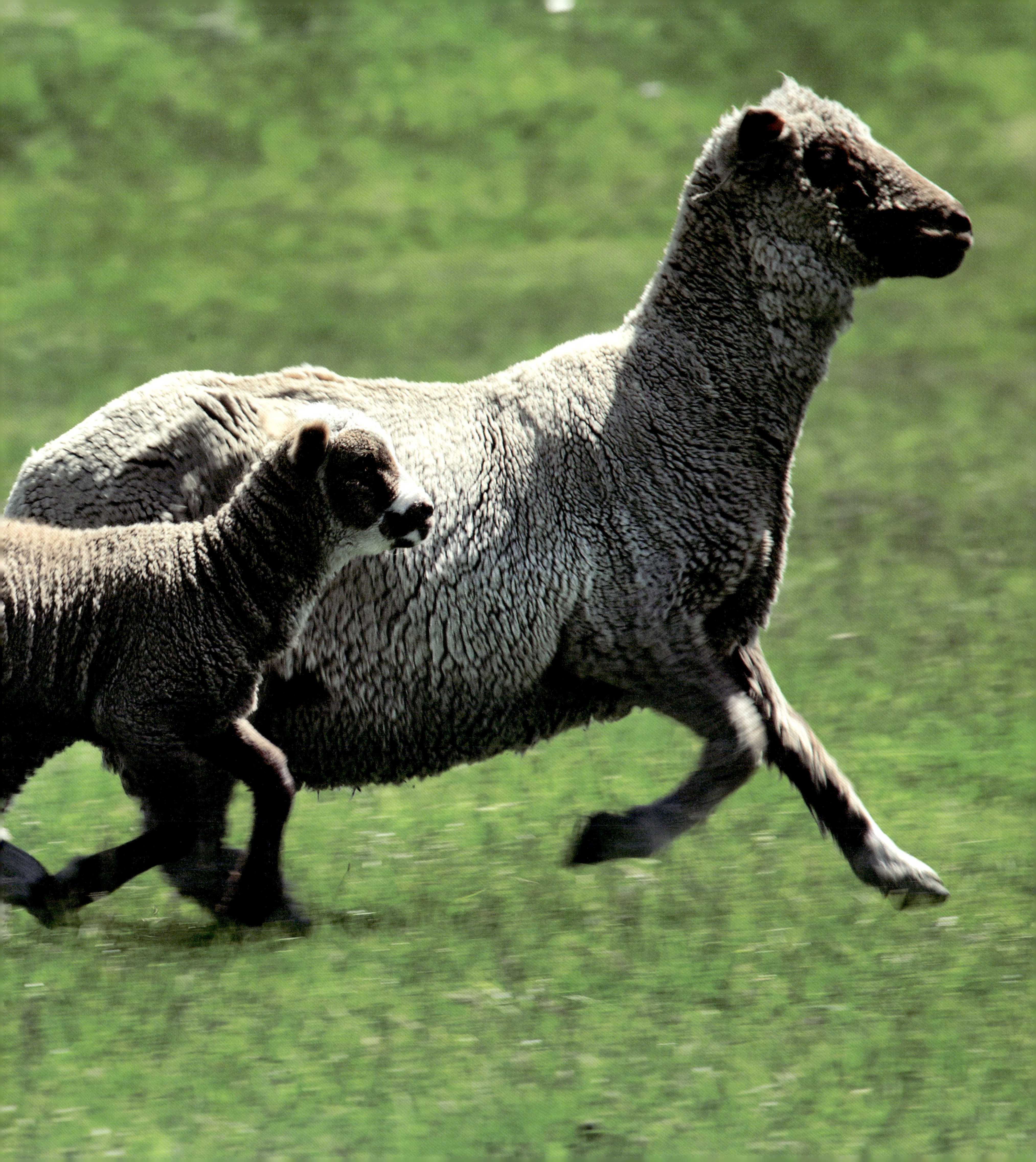

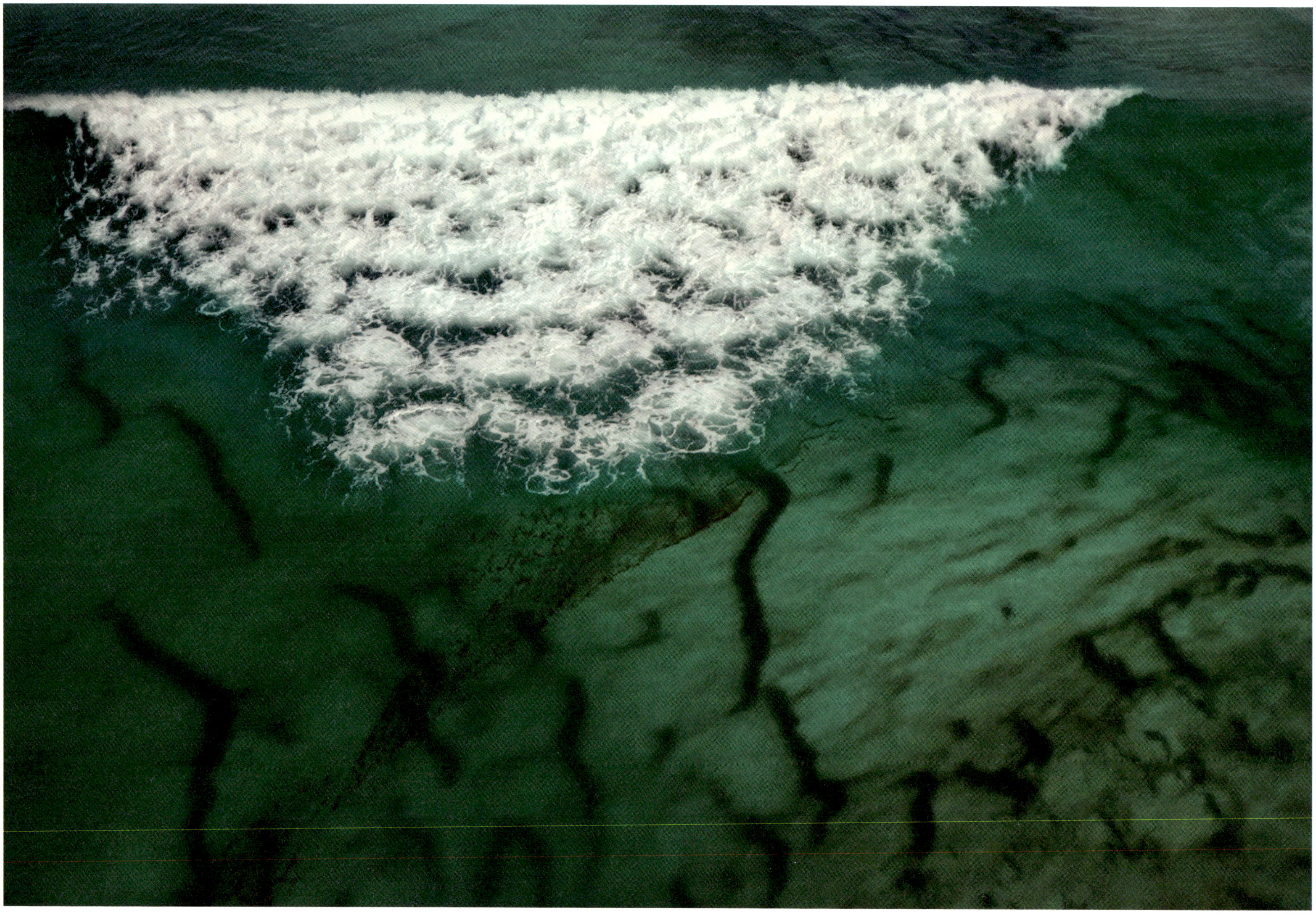

The aquamarine hues
off the Tasmanian coast.

Turquoise waters lap the rocks
around Mount Cook.

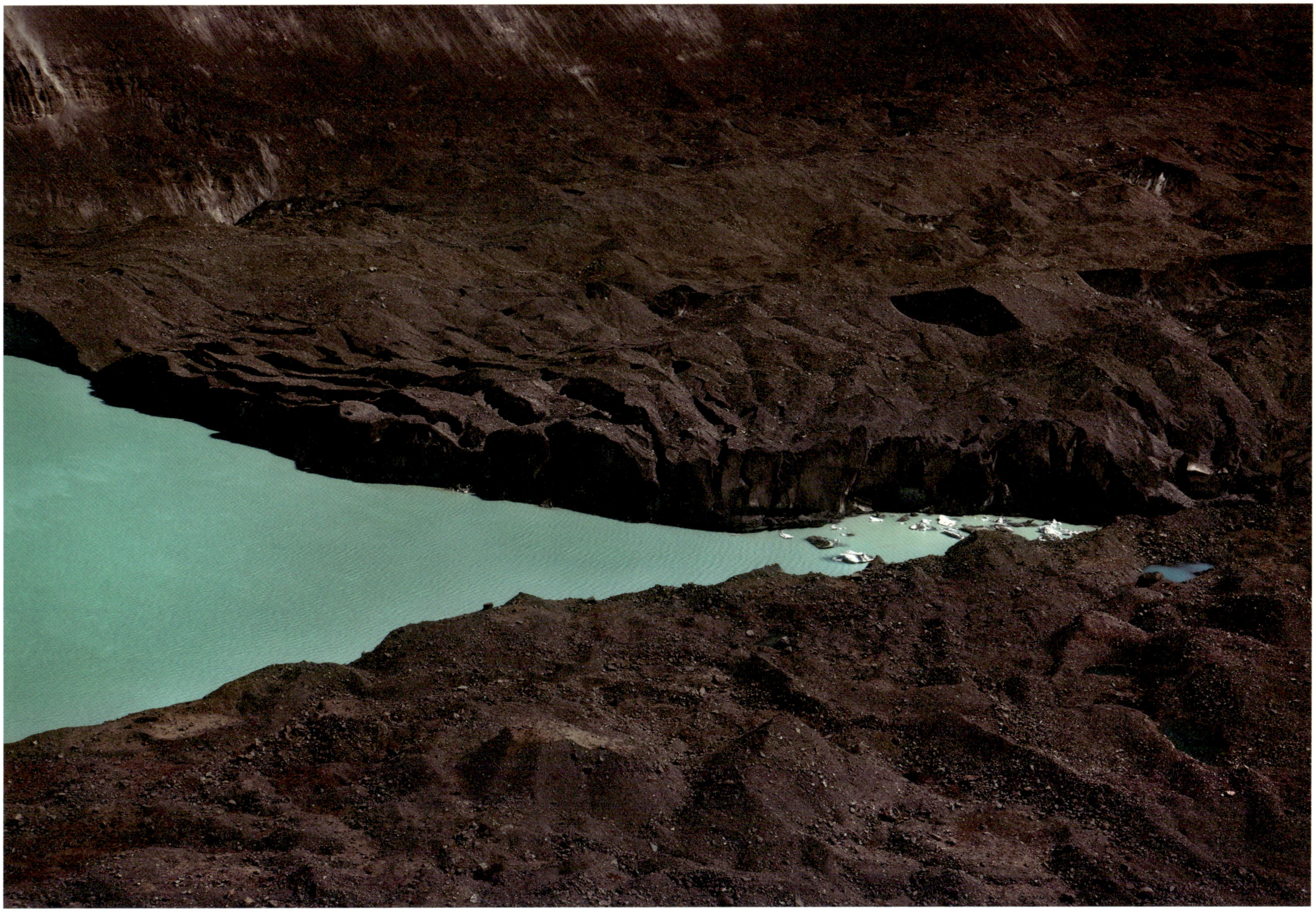

The dream cherished by these growers and by Loro Piana has become a reality thanks to the hard work that goes into every stage of the process: a combination of technical know-how, experience, historical research and innovation. It is still striking to see farms applying cutting-edge technology alongside traditional practices reminiscent of ancient rites.

On Visulea farm, New Zealand three generations of the Payne family are engaged in this craft; after the yearly shearing of the sheep, they hold a great celebration with food and music while they sort the fleeces. Each family member has a task and the most prestigious is that of selecting wool for Loro Piana's The Gift of Kings®.

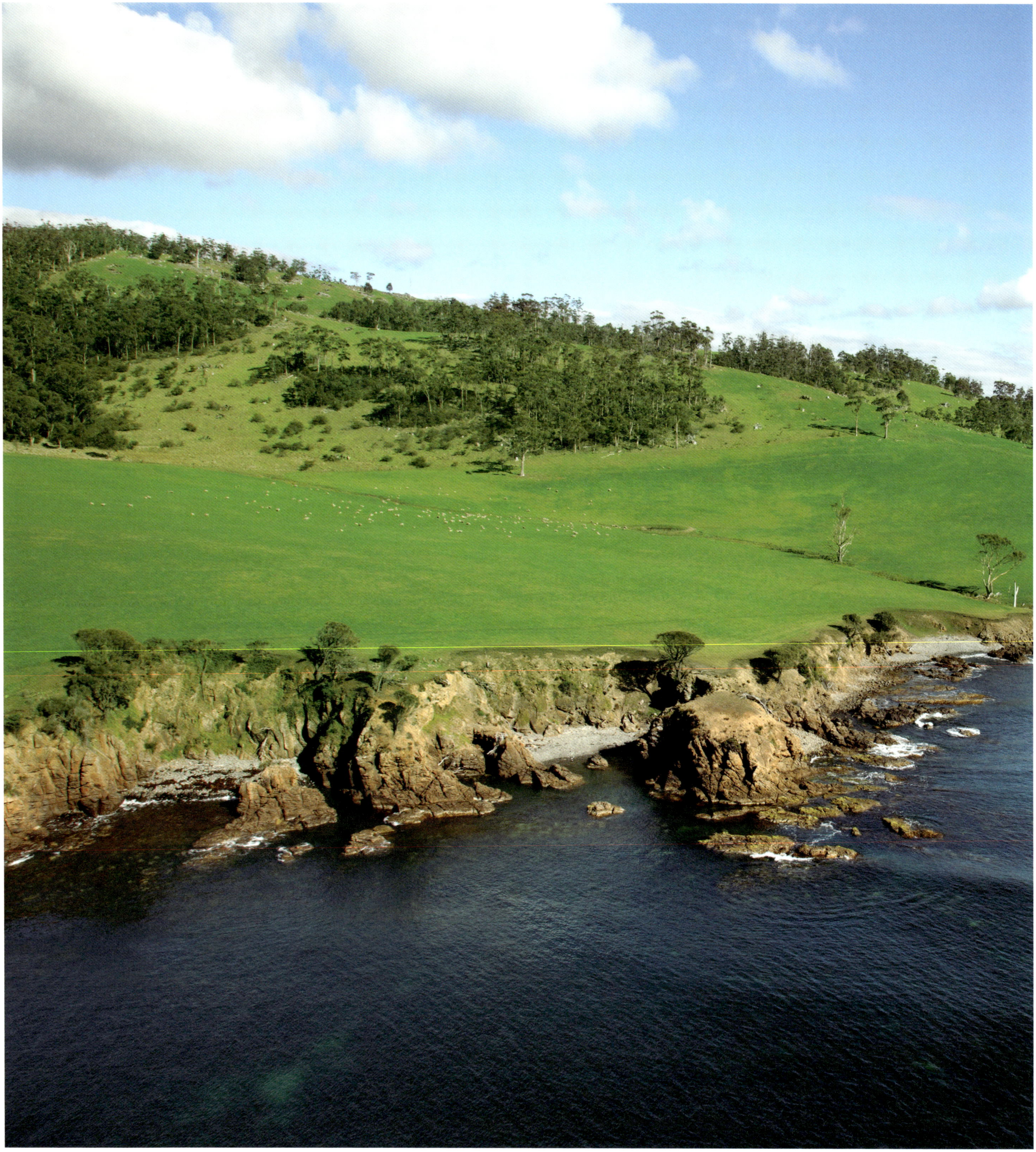

SHADES OF BLUE AND GREEN DOMINATE TASMANIA'S RUGGED COASTLINE.

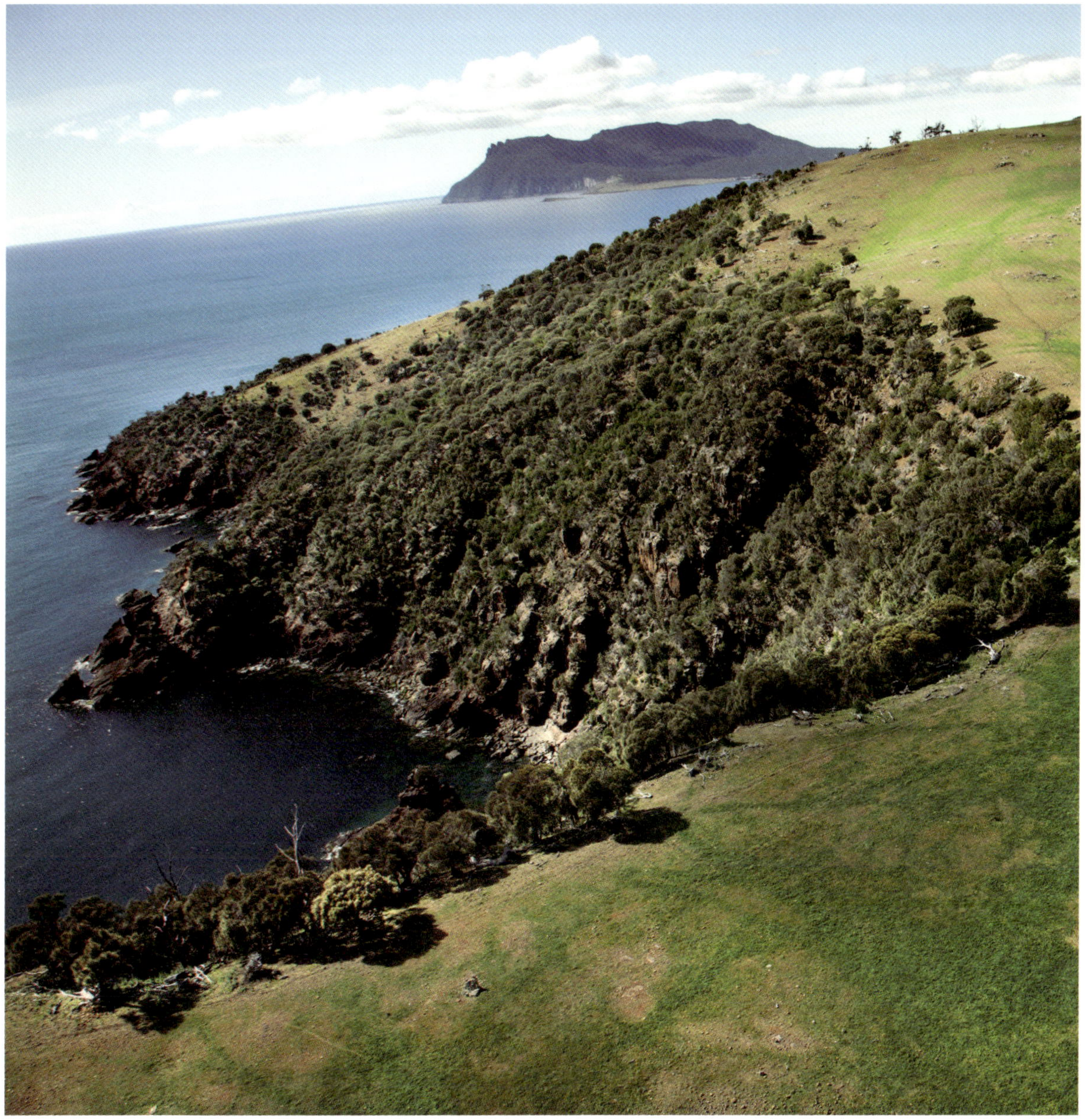

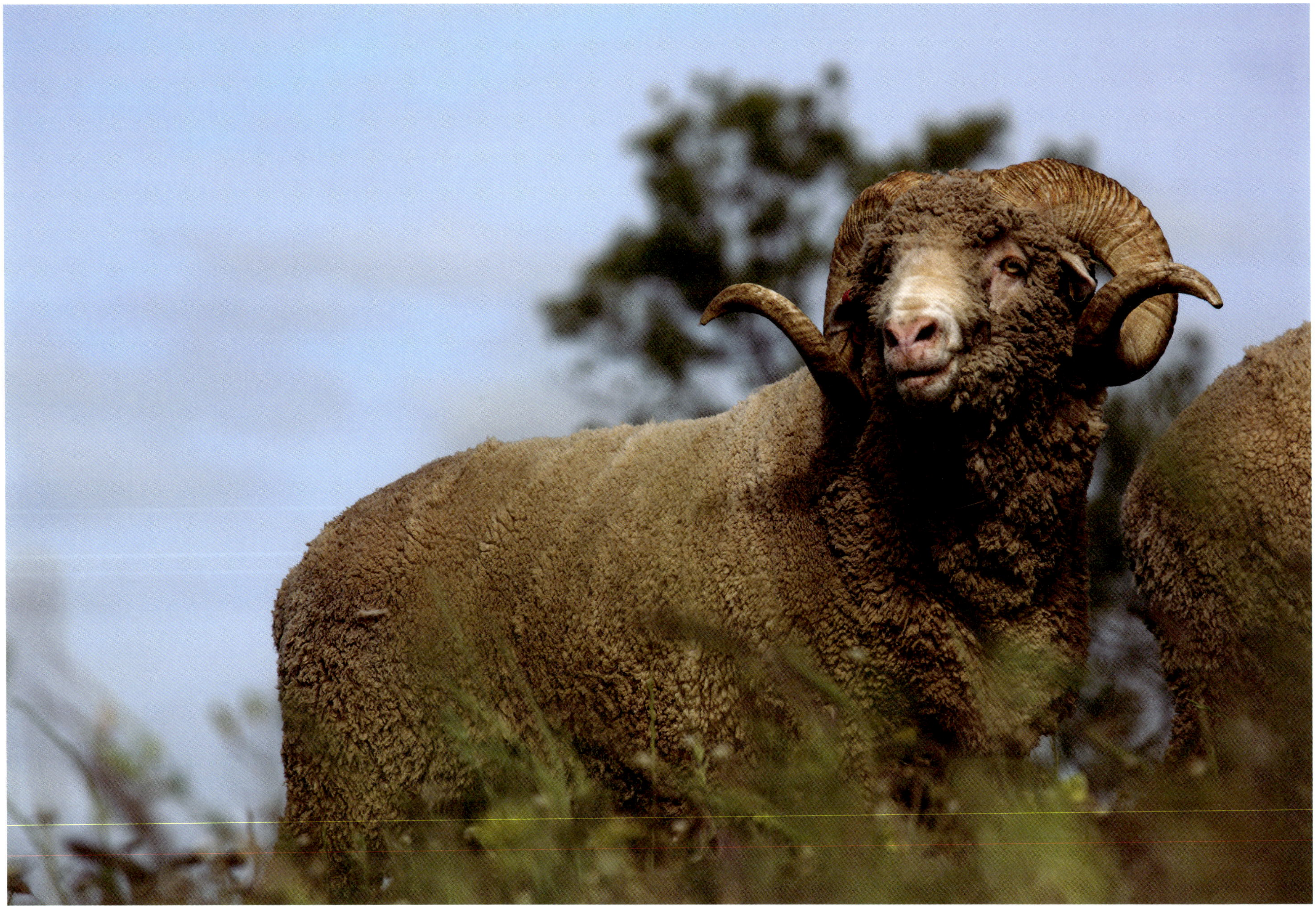

Merino sheep on Black Forest farm, New Zealand.

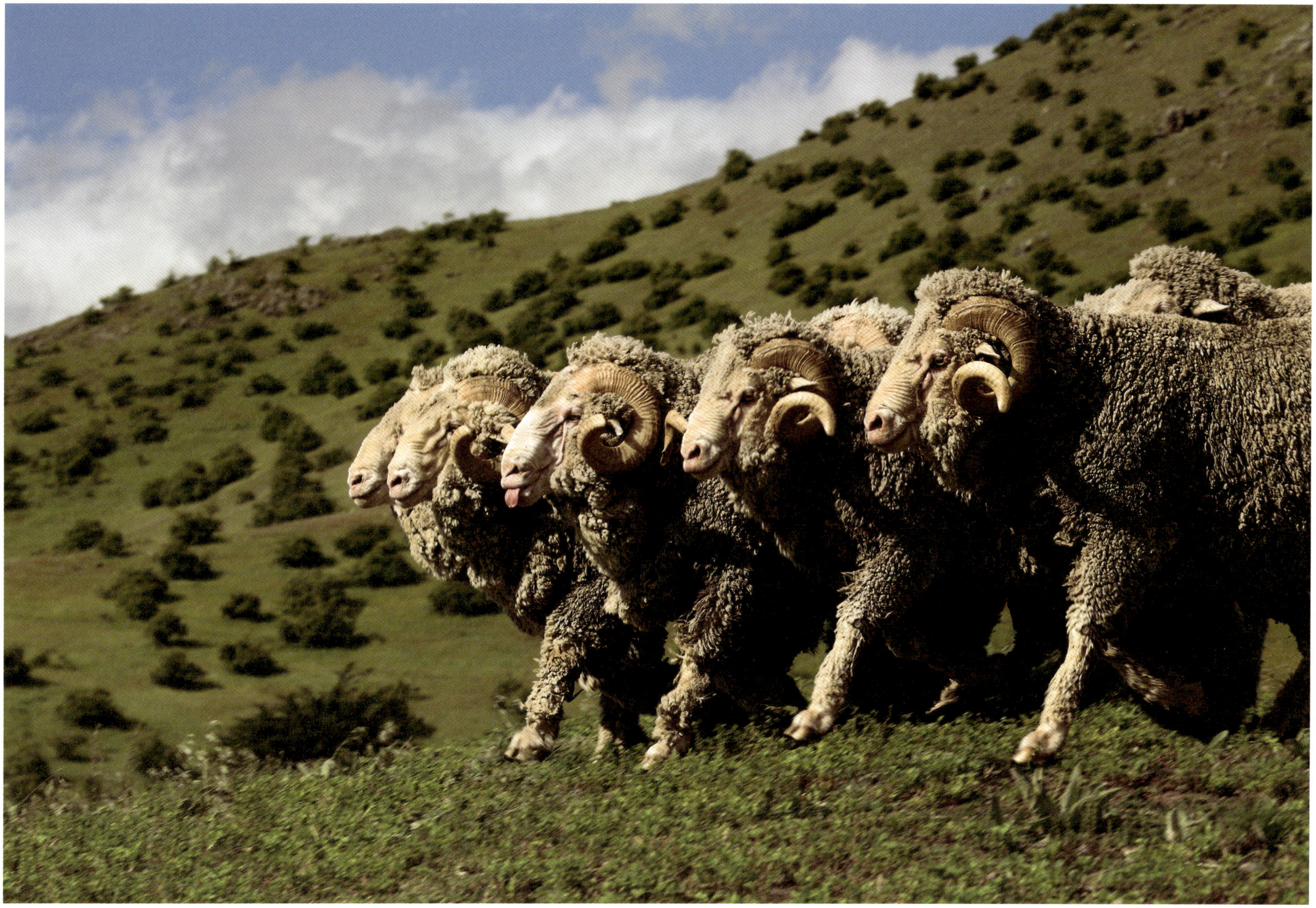

For the Paynes, working with Loro Piana means having the security to look to the future, make plans and progress. Like in an ancient court ceremony, the family consigns its finest wool to Loro Piana, which will present it to the world. Merino takes to the road once more, bringing people together and setting the seal on alliances and partnerships, just as it did in the days of the Spanish kings.

In the delicate balance that governs the ecosystem the merino sheep plays a discreet role: wool comes from nature and goes back to nature in a virtuous circle. As it is completely biodegradable, it is easily absorbed by the earth as a fertilizer.

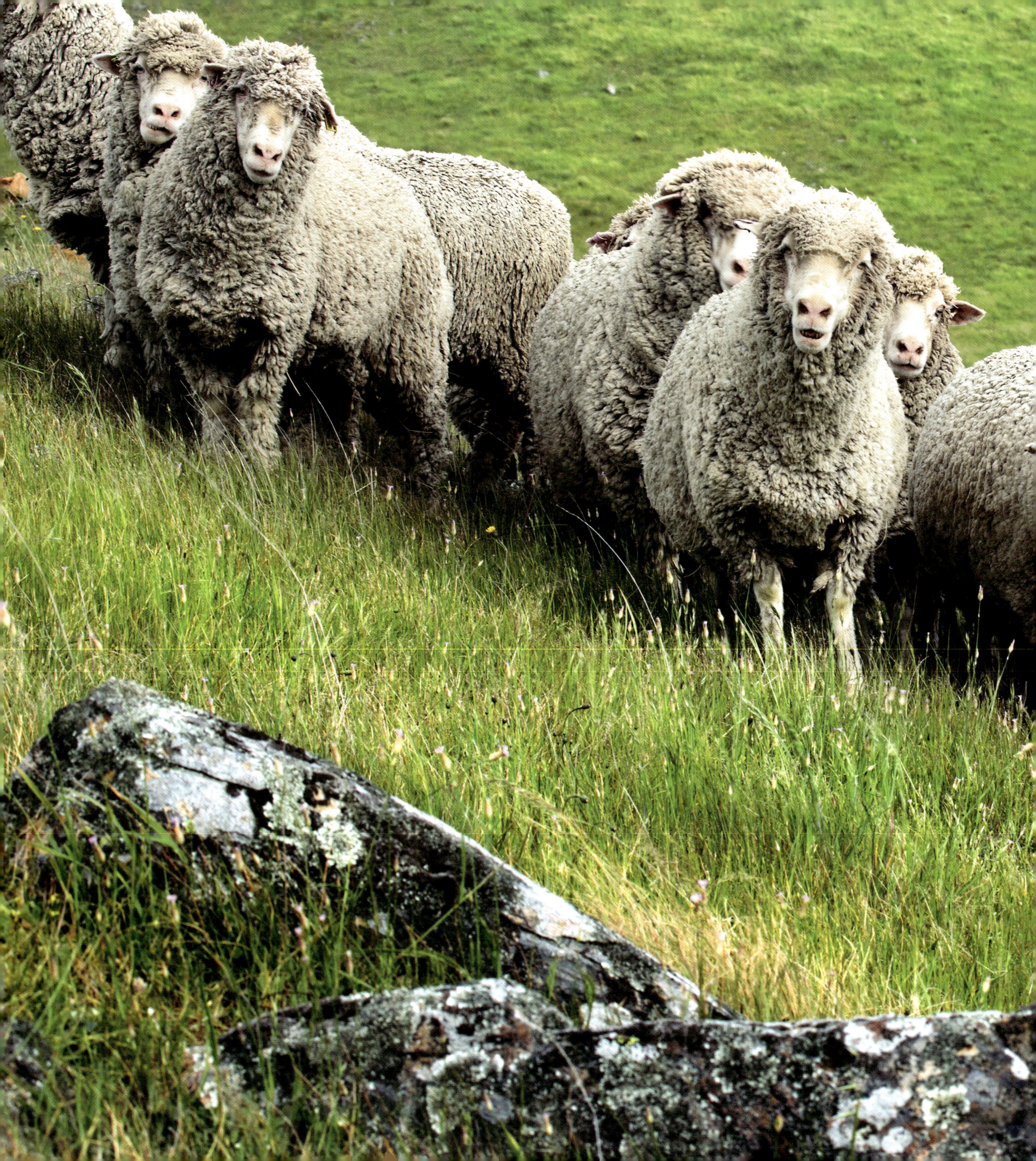

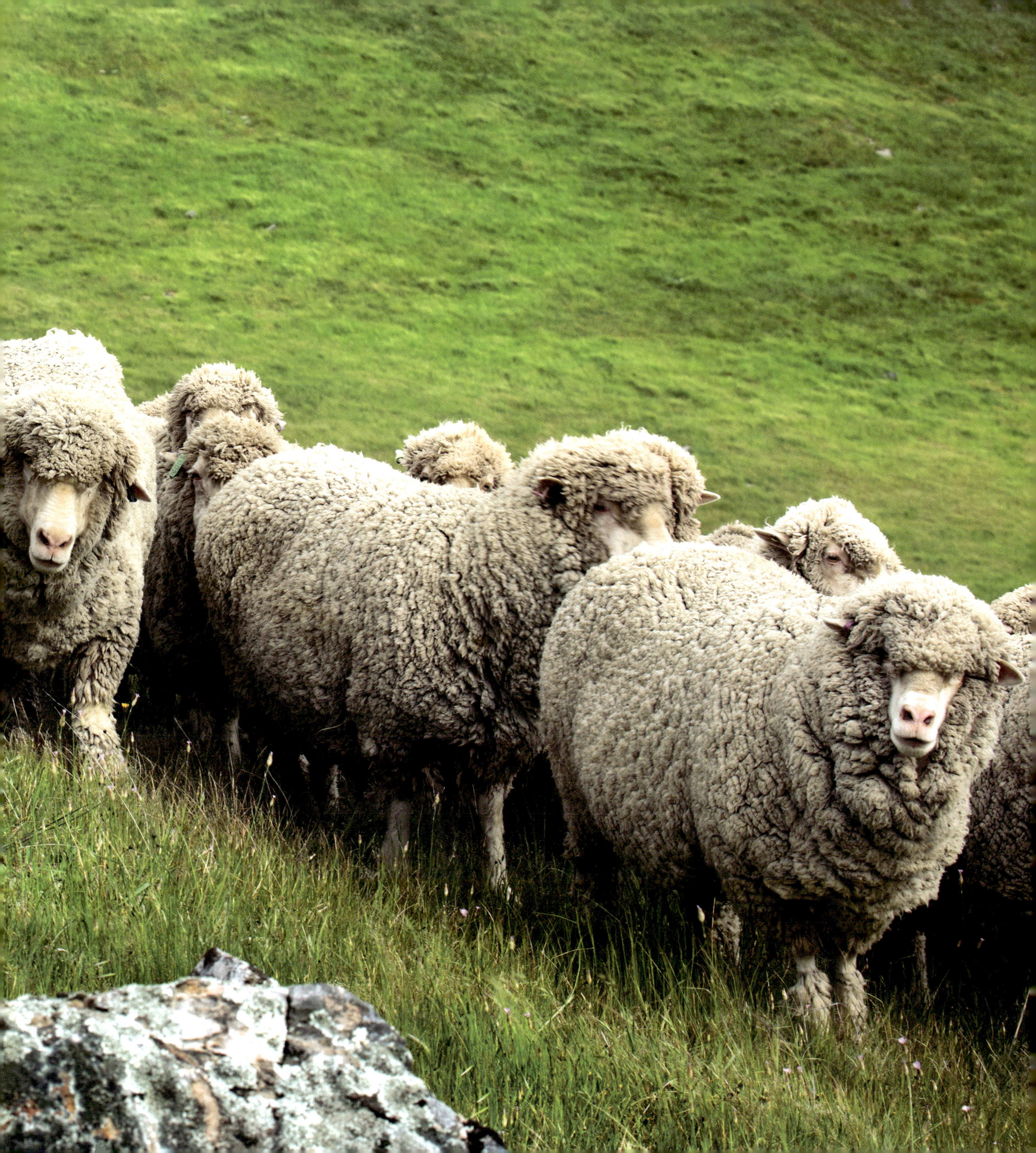

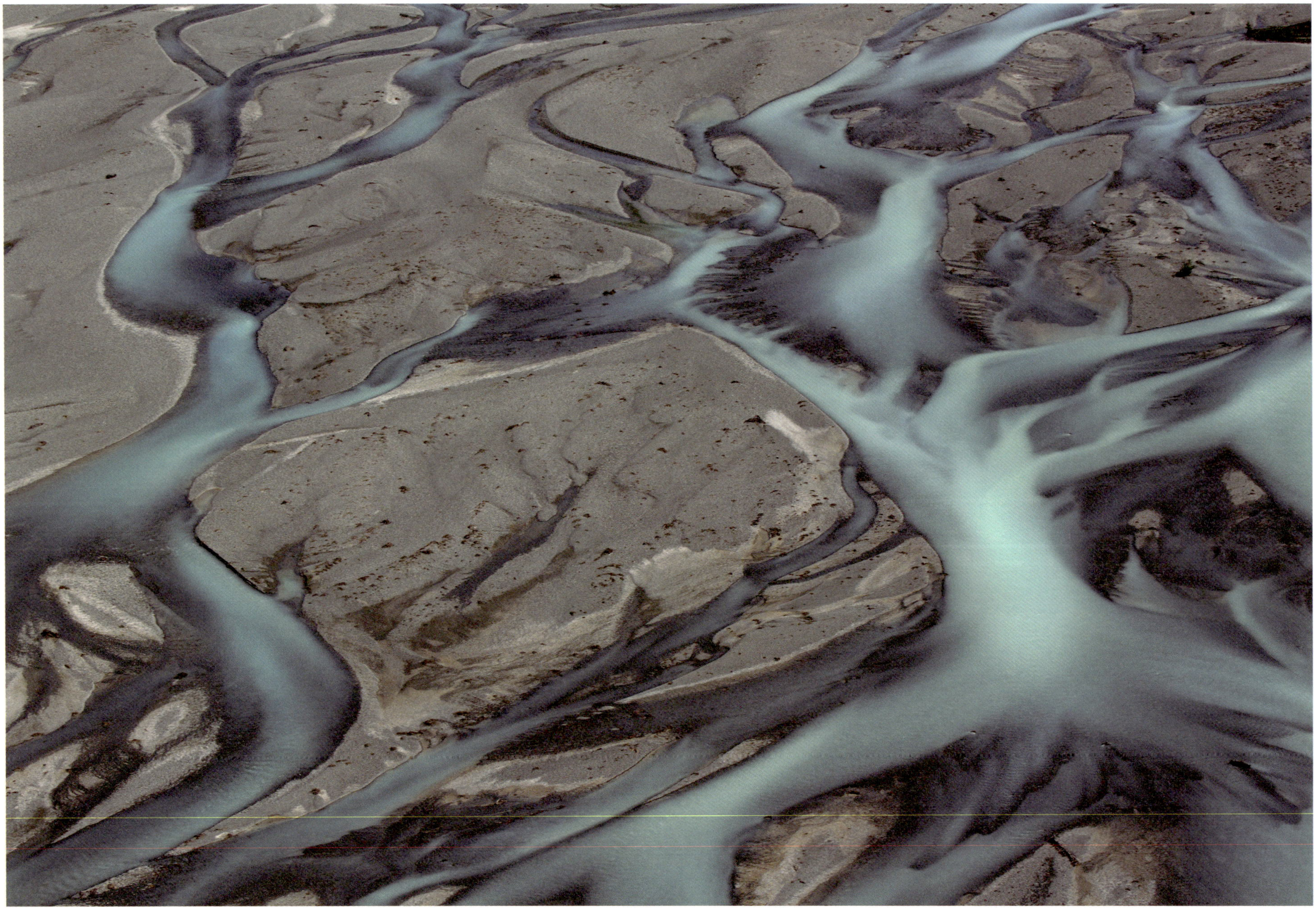

The magnificent lakes around Mount Cook
and Fiordland, New Zealand.

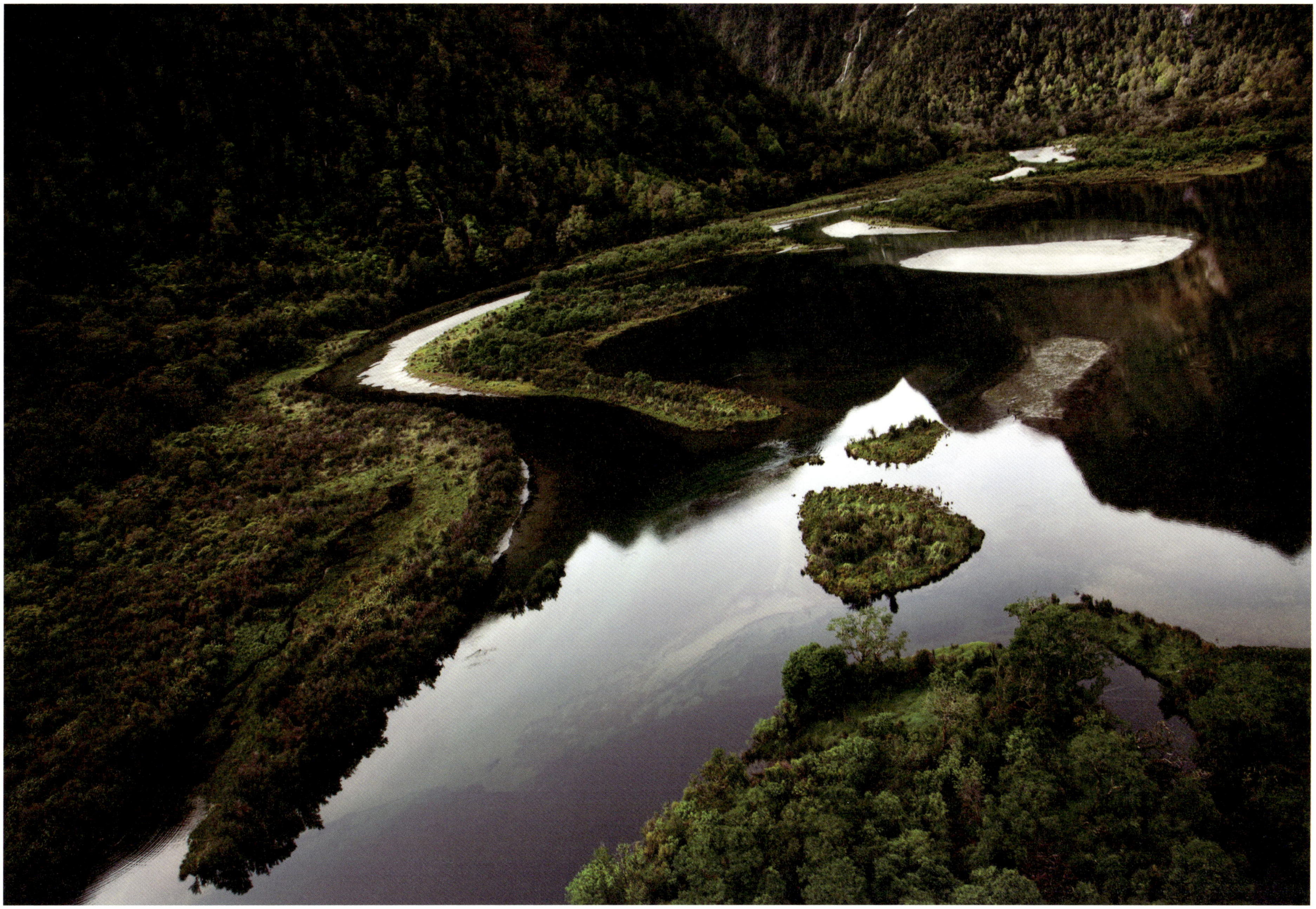

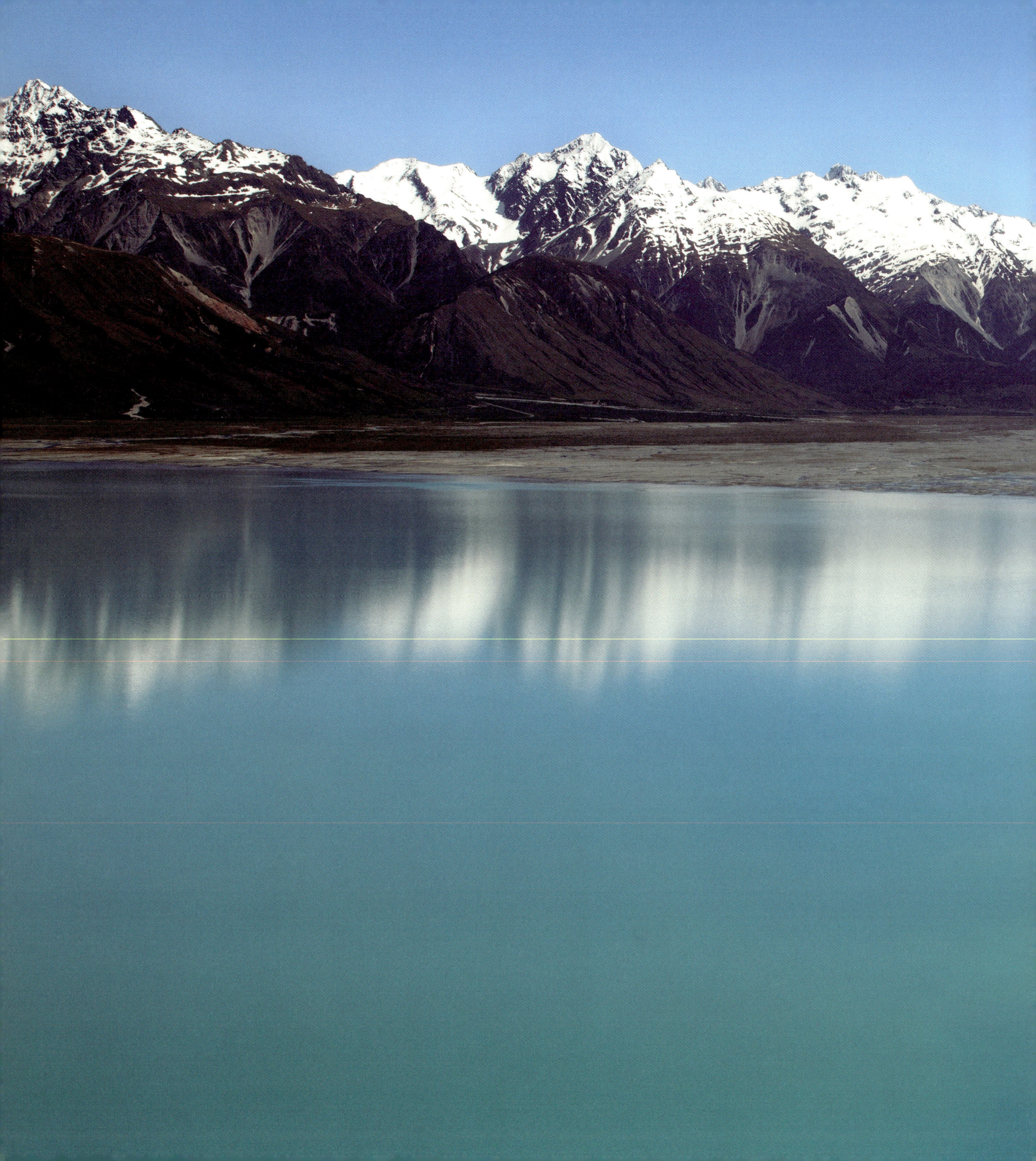

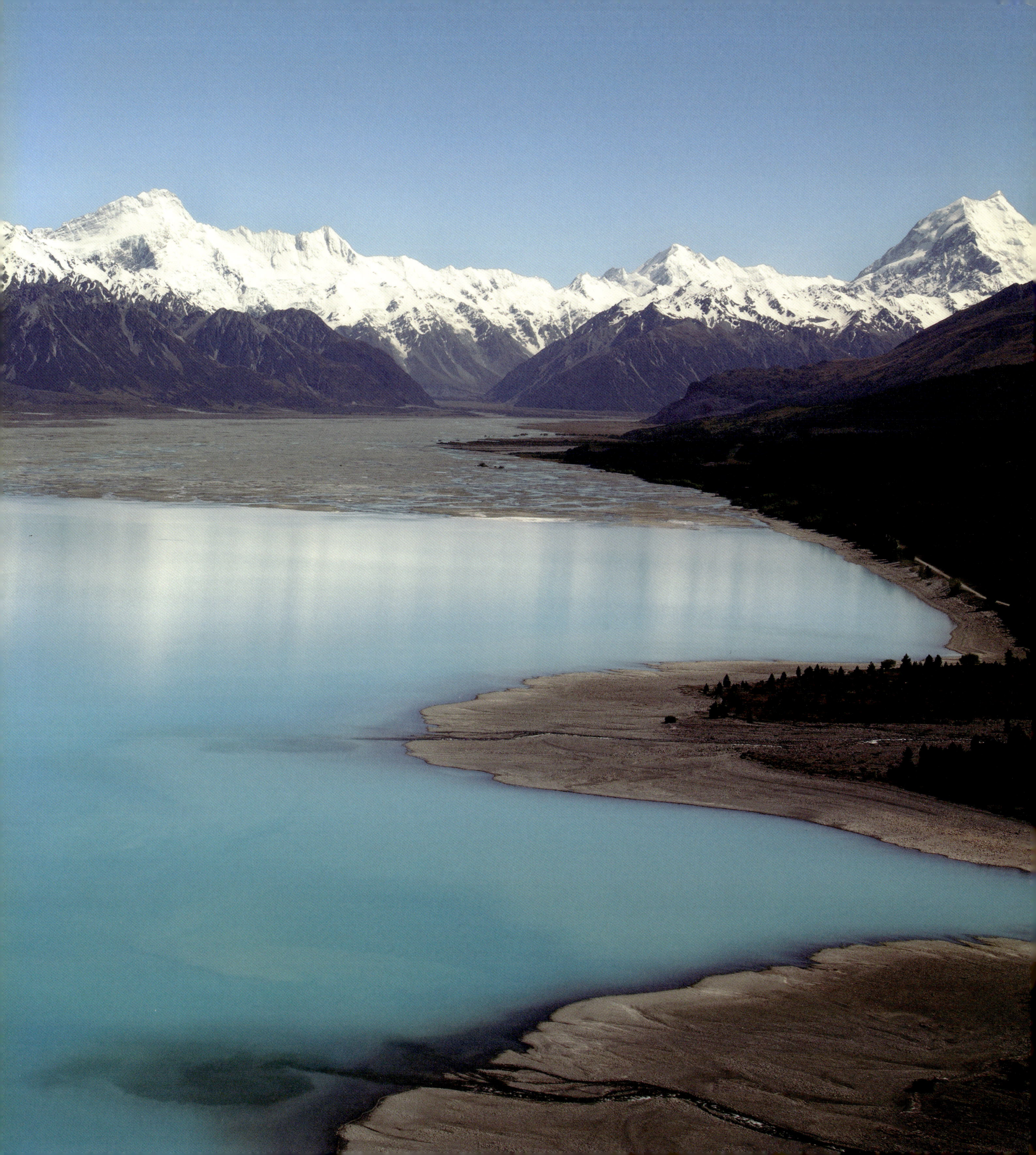

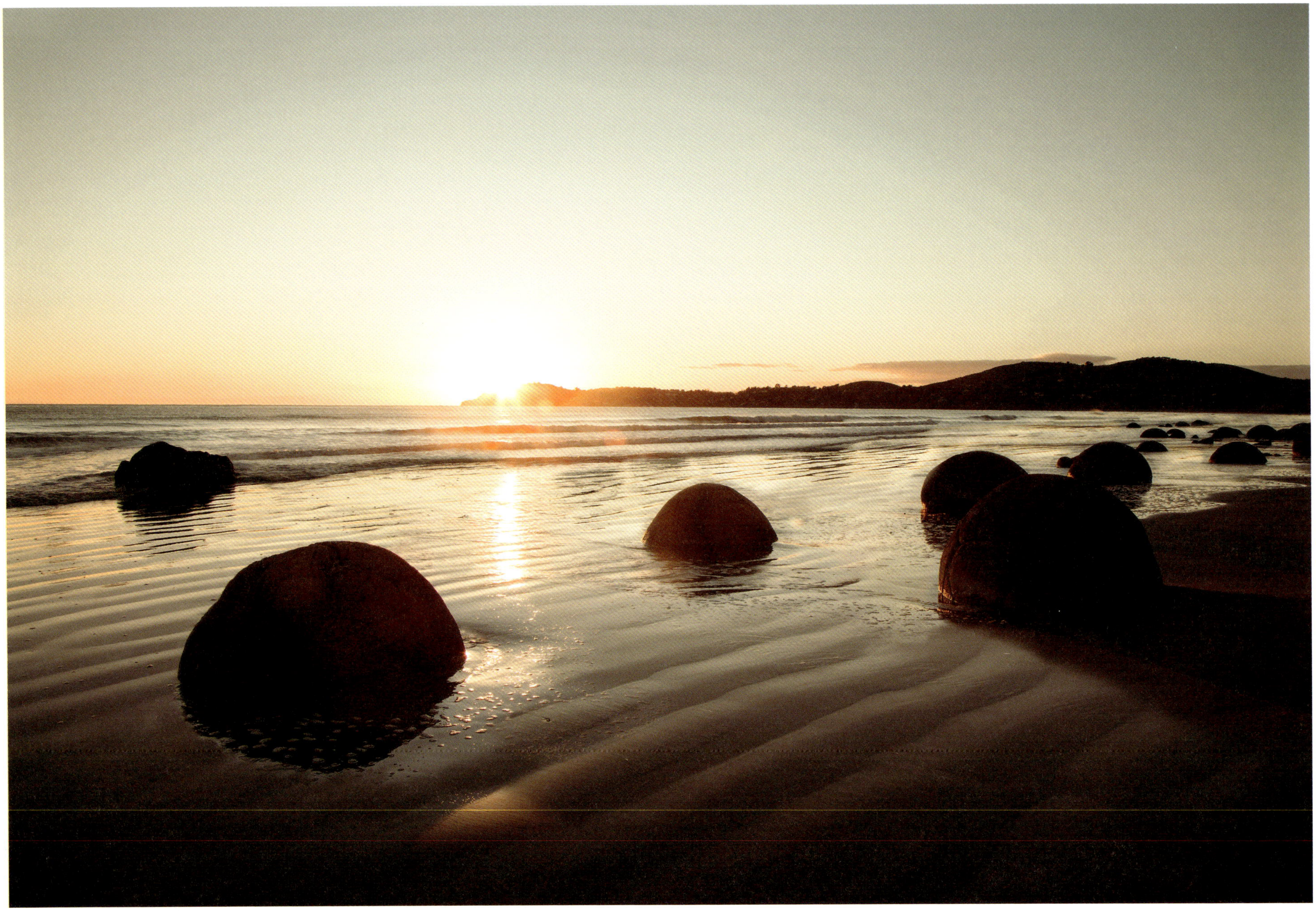

The characteristic Moeraki Boulders, huge round rocks strewn on New Zealand's beaches, look like giant stone bubbles emerging from the water.

previous pages
Mount Cook, New Zealand.

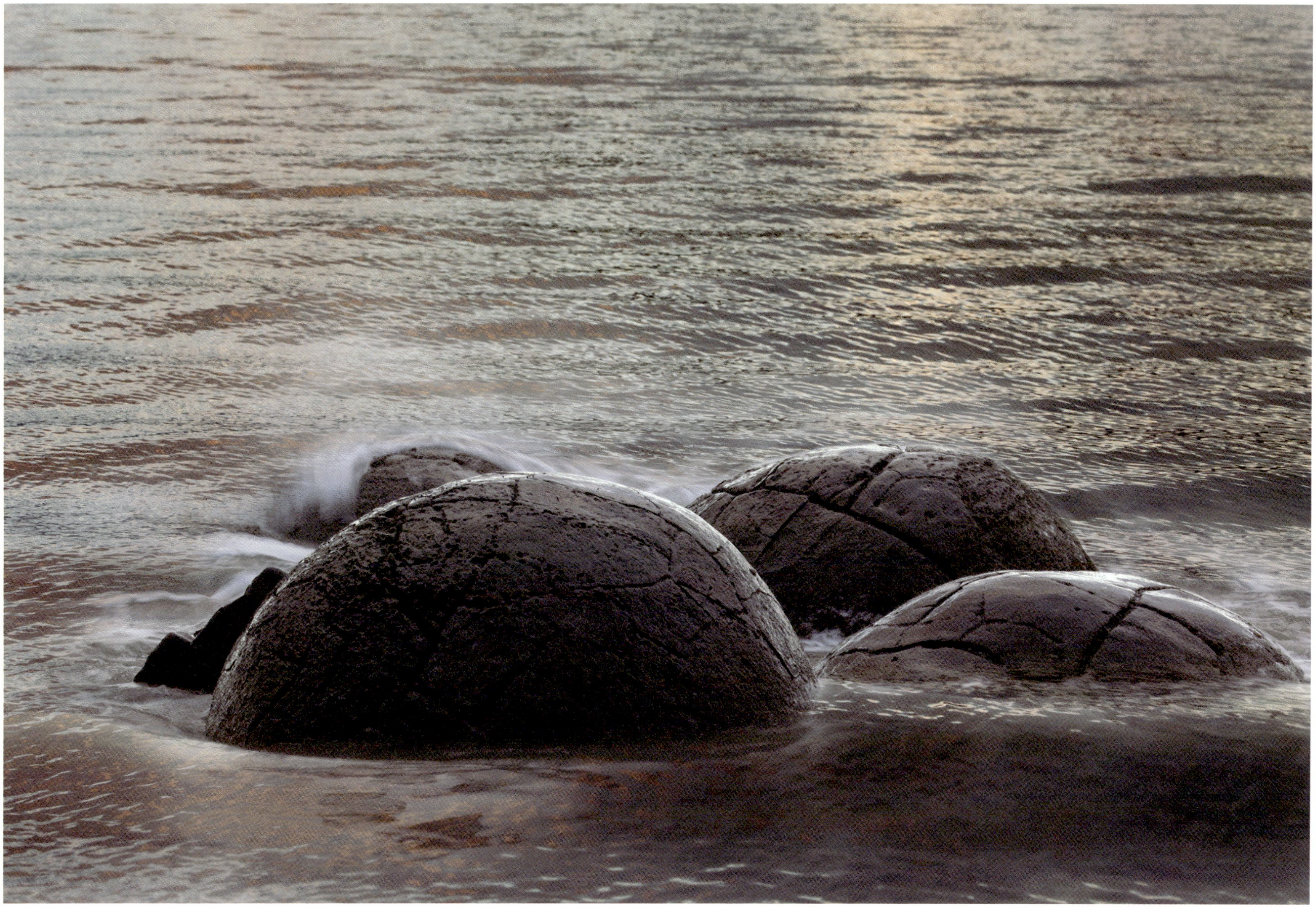

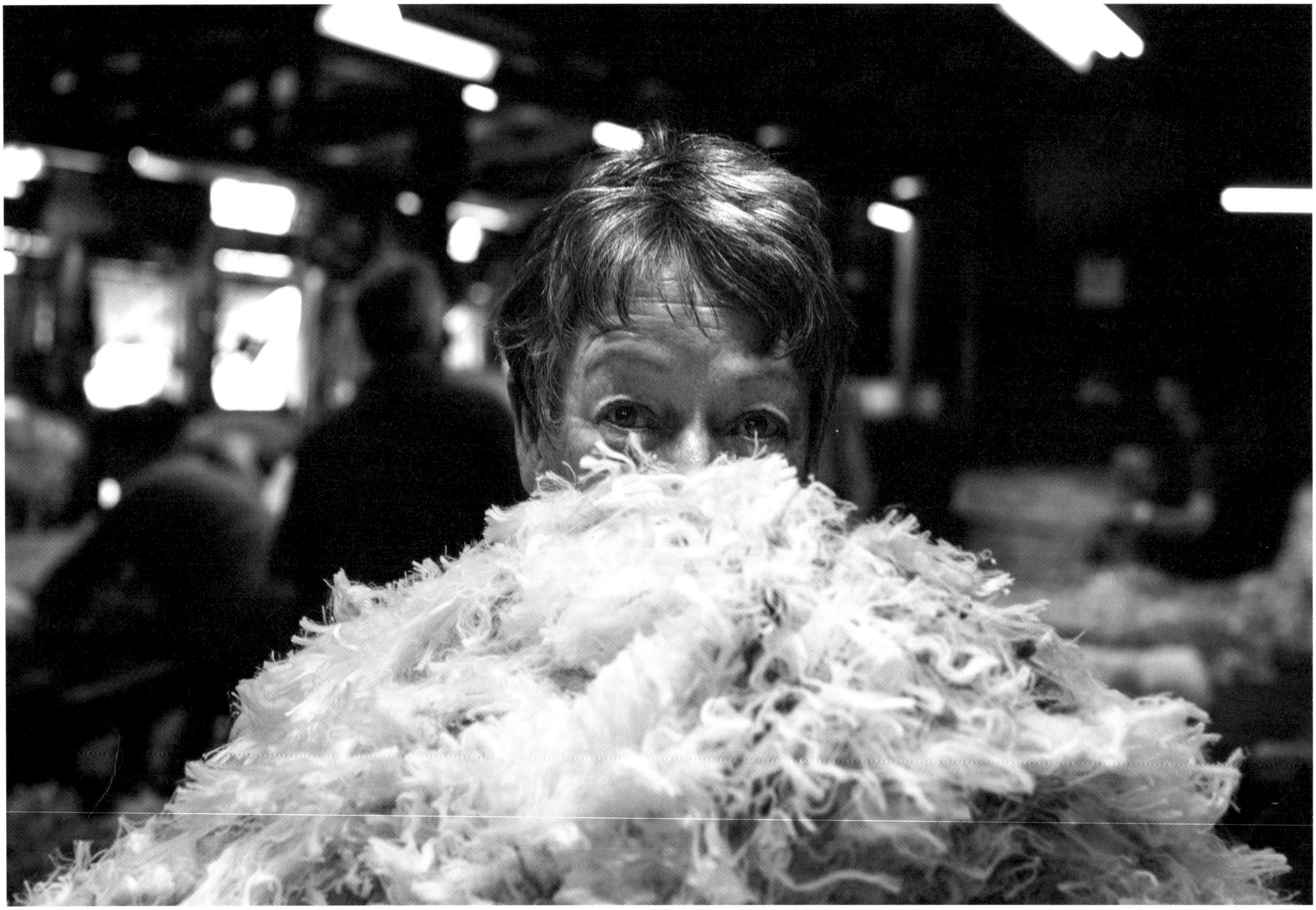

It is said that the best merino wool
has the scent of carnations.

following pages
The mountains of Fiordland,
New Zealand's largest national park.

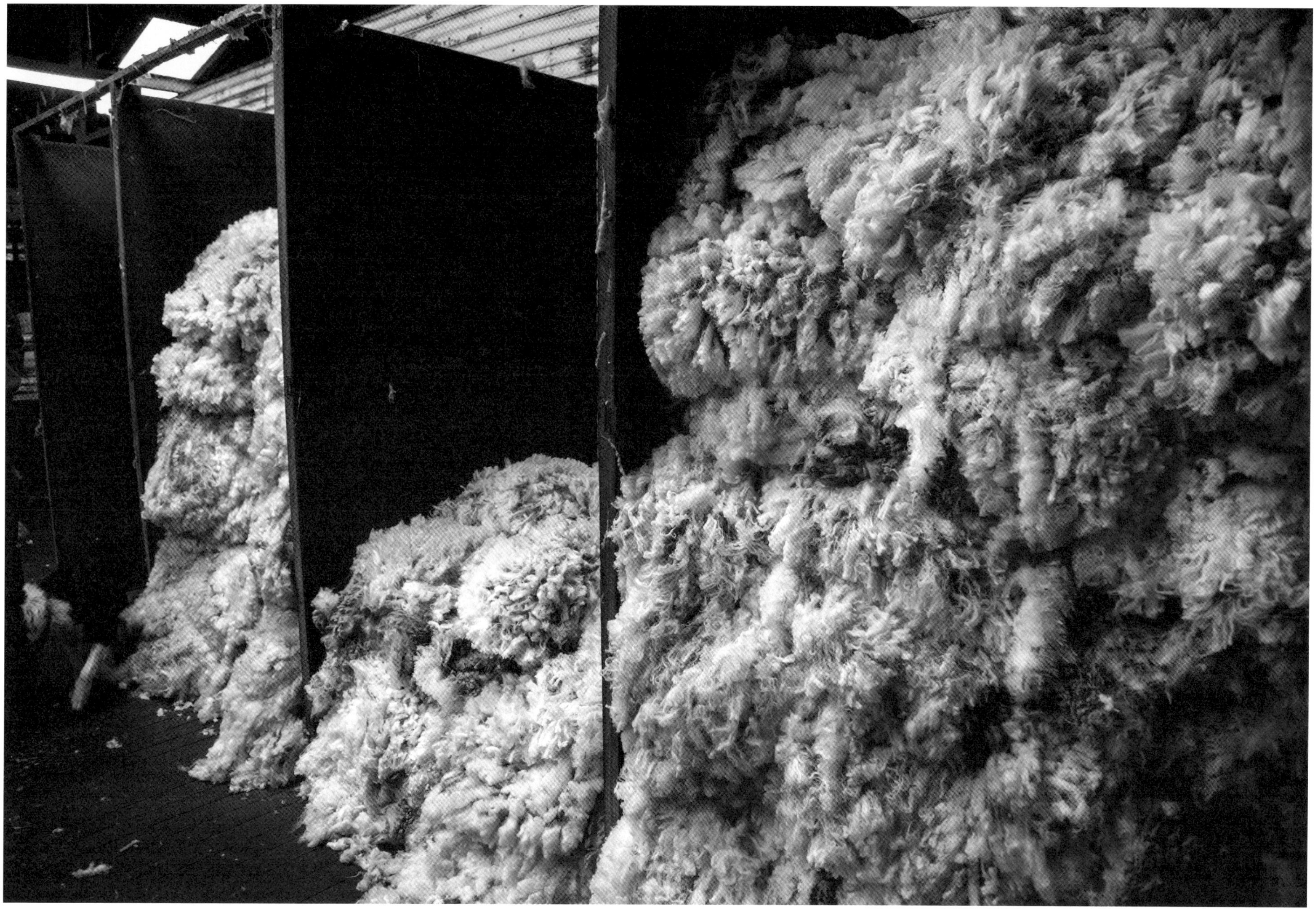

The Gift of Kings® is a rare and precious natural product that man has jealously guarded. Now, thanks to Loro Piana and the support it provides to growers, as well as to the specialized techniques they deploy, it has attained an outstanding standard of quality.

This noble task occupies entire communities that gravitate around magnificent farms in New Zealand and Australia, with whom Loro Piana has been working for years, supporting and encouraging their commitment to quality and excellence. Loro Piana has invested in something more than a sophisticated product; it has invested in long-lasting, dependable, fruitful relationships with those who, for generations, have cared for this animal. A gift for kings, a gift of nature.

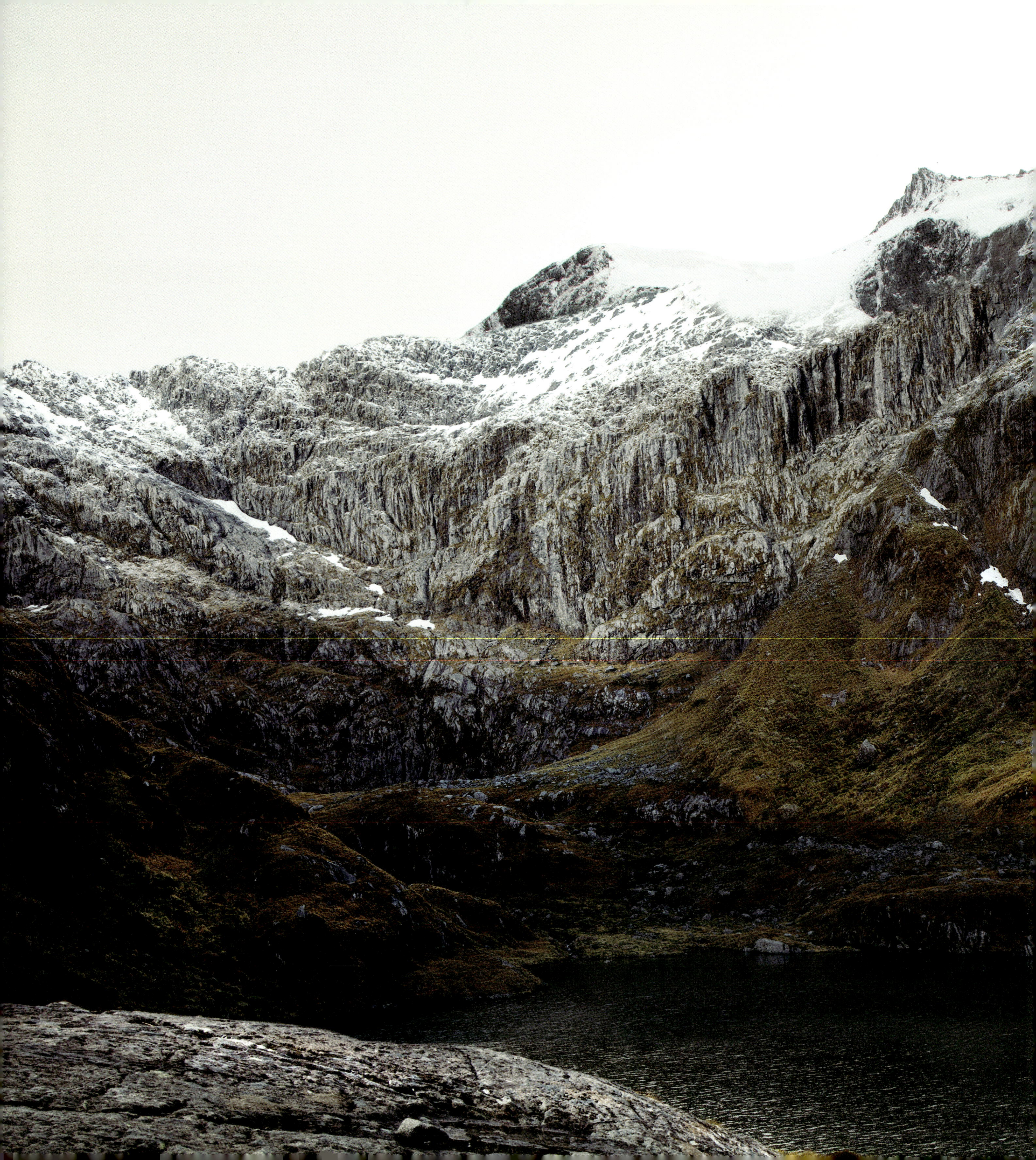

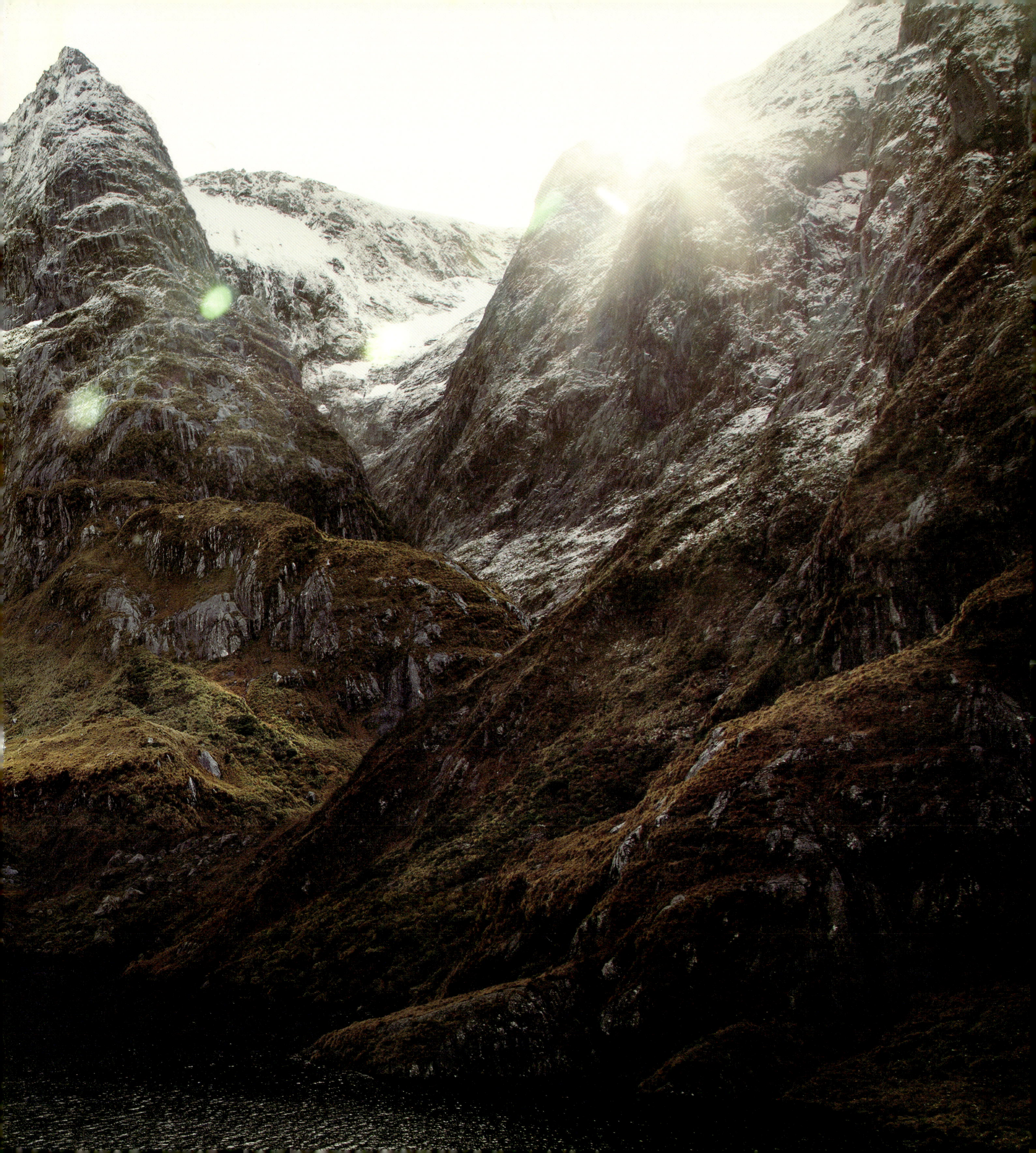

Edited by
Loro Piana

Texts by
Arianna Piazza

Design and Layout
Gritti Morlacchi & Schirmer *Milano, Venezia*

Editorial Coordination
Eva Vanzella

Copy Editor
Emanuela Di Lallo

Translation
Anna Carruthers

Photo Credits
© 2018 Bruna Rotunno

First published in Italy in 2018 by
Skira editore S.p.A.
Palazzo Casati Stampa
via Torino 61
20123 Milano
Italy
www.skira.net

Printed and bound in Italy. First edition

ISBN 978-88-572-3855-5

Distributed in USA, Canada, Central & South America
by ARTBOOK | D.A.P. 75 Broad Street Suite 630,
New York, NY 10004, USA.
Distributed elsewhere in the world by
Thames and Hudson Ltd., 181A High Holborn,
London WC1V 7QX, United Kingdom.